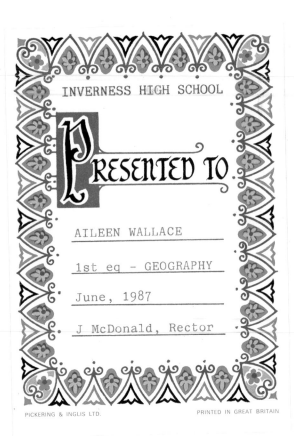

INVERNESS HIGH SCHOOL

Presented to

AILEEN WALLACE

1st eq - GEOGRAPHY

June, 1987

J McDonald, Rector

PICKERING & INGLIS LTD.   PRINTED IN GREAT BRITAIN

THE ENCYCLOPEDIA OF
WORLD
WILDLIFE

THE ENCYCLOPEDIA OF
WORLD WILDLIFE

*A survey of animals
and their habitats*

MARK CARWARDINE

*in association with
World Wildlife Fund*

OCTOPUS BOOKS

For my parents

First published 1986 by
Octopus Books Limited
59 Grosvenor Street
London W1X 9DA

ISBN 0 7064 2437 9

Printed in Hong Kong

CONTENTS

FOREWORD
by Sir David Attenborough, C.B.E., F.R.S.

The natural world today is in great peril. If it is to be protected against the increasing tide of destruction, then we all need not only to be aware of its beauties and wonders, but to understand the processes that have brought them into being and on which they depend. By introducing the world's varied habitats and the frightening problems of habitat destruction which are threatening so many species with extinction, I believe that this book will help to develop and encourage that understanding. At the same time, it is most generous of Mark Carwardine and his publishers to dedicate a proportion of its revenues to the World Wildlife Fund, one of the few organisations that works world-wide to protect animals and plants wherever they are threatened.

INTRODUCTION

No-one knows exactly how many different kinds of animals and plants live in the world today. Some say there are two million, some claim there are ten million or more. But one thing we do know is that many thousands – possibly hundreds of thousands – of these species are in serious danger of extinction, of being wiped out and lost for ever.

It is virtually certain that many species are disappearing even before we are aware of their existence, perhaps hidden away somewhere in a quiet corner of a tropical rain forest or in the depths of an unexplored sea. They will never even be named before they succumb to habitat destruction or some other threat to their survival. Among those which we know are in danger, many are so rare that only a handful of individuals are left.

There are only two surviving dusky seaside sparrows, both living in a zoo in Florida. Thousands of kilometres away, in Africa, there are only a dozen northern white rhinos, fighting for survival alongside elephants, leopards and all the other endangered species living on that beautiful continent. In Tasmania the thylacine, a strange dog-like animal related to kangaroos, has not been seen for about two years. Sadly many other animals and plants, in virtually every country in the world, are in similar trouble. One glance at the many volumes of *Red Data Books*, which are the authoritative source of information on the world's threatened species, gives some idea of the seriousness of the problem.

Extinction, of course, has been happening for millions of years. Plants and animals were becoming extinct long before people arrived in the world – and will continue to do so for millions of years to come. At least 90 per cent of all species that have ever existed have disappeared. In the past they went fairly slowly and as a result of natural processes, but since the arrival of modern man the extinction rate has soared. Scientists estimate that during the great extinction of the dinosaurs 65 million years ago, species were disappearing at the rate of about one every thousand years. Today, the extinction rate is more like 1,000 species each year.

Even common species are potentially at risk. The passenger pigeon, for example, was once so common in its native North America that it accounted for nearly 40 per cent of that continent's entire bird population. Enormous flocks, tens of kilometres long and several kilometres wide, were a common sight in the early 1800s. Indeed, scientists now believe that the passenger pigeon was once the commonest bird that ever lived on earth. Yet people managed to hunt it to extinction within just 50

Bat-eared fox This fox (*Otocyon megalotis*) can easily be recognised by its short legs and enormous ears. Found in East and parts of southern Africa, it prefers arid grasslands, savannas and bush country. It is the only member of the dog family to eat mostly insects, such as termites and dung beetles. These are most common around animals such as zebras and wildebeest, which is why bat-eared foxes are usually found near large herds of hoofed animals. The fox's ears are used to locate insects by sound – and they also act as radiators to help the animals lose heat during the hottest part of the day.

Oak bush cricket Bush crickets, or long-horned grasshoppers, are more nocturnal than true grasshoppers, becoming active in the late afternoon and continuing to 'sing' until well into the night. The sound they make is very high pitched and is usually produced by rubbing their two wings together. The oak bush cricket (*Meconema thalassinum*), however, is unusual in that it has wings which are as long as its body but they bear no 'teeth' for rubbing together. Nevertheless, it makes its own peculiar sound by beating on the ground with one hind leg and swinging its abdomen like a pendulum at the same time.

years. The last passenger pigeon seen in the wild was shot by a young boy on 24 March 1900; on 1 September 1914 the last member of the species died in captivity, in the Cincinnati Zoo.

How many more of the so-called 'common' species in this book will suffer a similar fate? The answer is probably 'a great many', because there is a danger many times more serious than hunting – the threat of habitat destruction. People are destroying habitats, the natural places where animals and plants live, at an even more frightening rate than they are exploiting individual species. And without habitats there can be no animals or plants.

Yet we continue to chop down tropical rain forests, turn rich soil into desert, drain important wetlands and destroy other habitats with complete disregard for the future. Common and endangered species alike will become extinct if this destruction continues at present rates.

Every habitat in this book is threatened by the actions of people. Some, such as the rain forests (which contain half the world's animal and plant species), could disappear entirely by the end of the century. Many rain-forest species are already threatened by hunting, pollution, competition with domestic animals and plants and a variety of other pressures – but habitat destruction is by far the greatest threat of all. Even such seemingly safe habitats as the vast oceans and isolated islands have not escaped. Ironically island species are among the worst-hit of all, and the populations of many ocean animals, including fish, whales and seals, are also very seriously depleted.

Some unthinking people ask: why conserve? Why should we bother about rare, obscure and apparently useless plants and animals? Well, a few moments' thought reveals the most obvious reason – that people will ultimately suffer. We rely on habitats, and the animals and plants that live in them, for our food, building materials, medicines and many other necessities of life. We also rely on the as-yet unstudied species for progress, which is vital if we are to resolve the current problems.

Wild plants and animals are essential for modern medicine, even if only as the starting materials for making drugs artificially. Quinine, used to prevent and treat malaria; reserpine, used in the treatment of heart and circulatory diseases; snake poisons, used in non-addictive pain killers; and bee-sting venoms, used in the treatment of arthritis, are just some of the many medical products derived from wild animals and plants. Yet only a minute proportion of species has been investigated for

Desert lizard This lizard (*Agama mutabilis*) belongs to a group of brightly-coloured lizards, known as the agamids, which live in Europe, Africa, Asia and Australia. They are the Old World counterparts of iguanas, which live mainly in North, Central and South America. See also page 44.

Scorpion fish These fish (*Pterois* sp.) look deceptively harmless as they sit on rocks and coral reefs. But their beautiful spines can sting and cause agonising injuries. They live mostly in the Indo-Pacific Ocean but have been found considerably further afield. See also page 134.

their possible usefulness. Native peoples such as the Aboriginals in Australia, the Masai in East Africa and the Indians in South America are using thousands of local plant and animal types that have never been studied scientifically.

The world is believed to support about 80,000 edible (to humans) plant species. Less than 20 of these produce 90 per cent of all the world's food. Even if we restrict ourselves to established crops in the future, their productivity cannot be maintained without fresh variations from their wild relatives. These will help the plant breeders tackle new diseases, problems of drought and many other unforeseeable hazards in the future. In addition many possible alternative foods such as algae, seaweed and yams show great potential and may become more suitable as staple crops.

There are many other good, scientific reasons for protecting animals, plants and their habitats. But they also deserve to be conserved because they are interesting, attractive, inspiring – and even because it is questionable whether people have the right to exterminate a species or shift the course of evolution.

Since it was established in 1961, the World Wildlife Fund has invested over £45 million in conservation projects in more than 135 different countries. During the past few years, it has been turning its attentions more and more to the fundamental problem of habitat destruction.

The aim of this book is to demonstrate the importance of our planet's habitats and the wealth of wildlife living in them. Unless many more people learn to appreciate and respect the natural world, there will be no chance of preventing further and irrevocable damage. In this, the World Wildlife Fund's 25th Anniversary Year, we must therefore make special efforts to reverse the disastrous trends of the past.

Swallowtail butterfly The swallowtail (*Papilio machaon*) occurs throughout Europe but, in Britain, is found only in a few fens in the Norfolk Broads. The continental races are found in many other habitats, including mountainsides. The adult has a wingspan of up to ten centimetres, making it the largest butterfly in Britain. It lives for about 30 days – but many caterpillars and chrysalises live for only a very short time before being eaten by spiders, birds and small mammals. Swallowtails have disappeared from many parts of their former range because of wetland drainage.

How to use this book

In addition to the main text, further information on the species and habitats in this book is given in the form of maps and Data panels. World maps like the one on this page show the broad world distribution of each habitat. In the case of islands and cities, some of the most important examples are pinpointed and named. Because the polar regions are covered, the Antarctic continent is shown as an inset in the lower left-hand corner of each map.

The Data panels give specific information on each main species: scientific name, how it is classified by scientists, its distribution worldwide, the habitats in which it is most often found, its approximate size, and its preferred food. If the species is particularly rare, or threatened with extinction, it is labelled 'ENDANGERED'. Strictly speaking, 'endangered' is a specific term used to describe those species which are *most* in danger of extinction. But here it is used in its most general sense, for any species listed in the IUCN's Red Data Books, and therefore includes all those officially classified in the threatened, vulnerable and other categories.

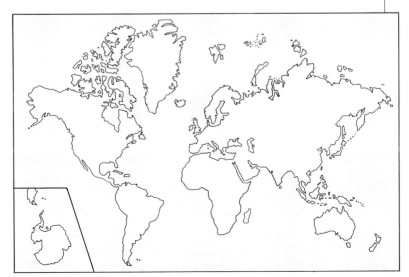

TROPICAL RAIN FORESTS

Tropical rain forests, or 'jungles' as we sometimes call them, are among the richest wild places on earth. Nowhere else in the world supports such a teeming variety of animals and plants or hides the solutions to so many mysteries.

From the highest treetop to the darkest forest floor, the rain forests are alive with plants and animals – clouded leopards, gorillas, orang-utans, tapirs, parrots, frogs, birdwing butterflies, and literally hundreds of thousands of others. A single hectare (two and a half acres) of forest can contain up to 200 different species of trees, compared with only 10 in the same area of typical British woodland; and it is possible to find as many different bird species in one hectare of tropical rain forest as have ever been recorded in the whole of the British Isles.

Not surprisingly, these forests contain at least half the world's species of animals and plants. There are many more which have never been seen or identified by man. Yet they cover only about one-tenth of our planet's land surface, restricted to a narrow broken belt straddling the Equator in three main regions: South America, Africa and South-East Asia.

Types of rain forest

There are many different kinds of tropical rain forest: mangrove swamps along the coasts, dense forests of the lowlands, and montane (mountain) or 'cloud' forests on the uplands. In each type of forest there are well-defined layers providing a range of habitats at different heights above the ground. At the very top, perhaps 45 metres (150 ft) above the forest floor, is the emergent layer. Here the giant trees of the jungle thrust through the canopy and spread out well above their neighbours. They enjoy the first share of the sunlight, but have to pay the price of high temperatures, low humidities and strong winds. Directly below the emergent layer is the canopy itself, a far more hospitable place to live and the site of most animal activity. It acts as an umbrella over all the other layers, intercepting both the sun and the rain, and makes the forest beneath dark and steamy. Only about two per cent of the sunlight falling on the canopy reaches the jungle floor; some arrives as tiny flecks of bright light shafting through the gloom, the rest as pale rays of greenish light which has passed through the leaves high up above.

Passion flowers *above*
The bright colours of passion flowers (*Passiflora vitifolia*) are found in the rain forest of Costa Rica.

Rain forest canopy *right*
The dense, green canopy of this rain forest in Indonesia gives an idea of the thickness of the vegetation.

Human visitors to jungles, however, usually see very little animal activity. They normally view the forest from its edge (a river bank or road), where light-loving plants form a dense and virtually impenetrable mass. Those who do venture through this 'curtain' into the interior find it open and easy to walk through, but while they may expect to be confronted by man-eating plants, fierce big cats, snakes, bloodsucking insects and poisonous spiders, they are more likely to hear only a few strange sounds and see virtually nothing. The animals are there, but most are either nocturnal or seldom venture down from their arboreal homes, high in the branches. Their presence is betrayed only by faint rustlings in the leaves, and perhaps the distant crashing of monkeys through the foliage or the shrieks and calls of birds high up in the canopy.

To make the most of all these riches, rain-forest animals have developed many specialized features. Some, including various frogs, snakes, lizards and mammals, have developed the ability to glide through the air, in order to speed up their movements through the forest canopy. Monkeys, anteaters, pangolins and even a species of porcupine have prehensile tails to help them climb. Others, such as the jungle cats, are camouflaged with spotted coats that blend in with the dappled light of the forest floor.

Forest flora

Ultimately, of course, all these animals depend on the jungle plants for their livelihood. The steamy heat is ideal for many plant species, including brightly-coloured orchids, spectacular climbers and ferns, while the constant climate means that there is an ever-growing evergreen tangle of plant life all year round. There are flowers in bloom at all seasons – attracting bats as well as birds to their sweet nectar and energy-rich pollen – and there is always a ready supply of leaves and fruit. Even on the gloomy forest floor animals are able to take advantage of this food, which falls from the canopy above.

The forests and mankind

Unfortunately, in recent years there have been more human visitors to tropical rain forests than ever before. The jungles have always been home to ancient tribes, who for centuries have been chopping down trees for firewood and building materials, hunting animals, picking fruit and berries, and catching fish in the rivers. But people from the towns and cities are now rapidly infiltrating rain forests, bringing with them bulldozers, mechanical saws and cattle.

The forests provide many products which are very important to us. Every time we drink a cup of coffee, eat chocolate, peel a banana, crack Christmas nuts or use anything made of natural rubber, we are using rain-forest products. The plants and animals provide many important life-saving drugs used in medicine, including quinine and curare, and we use the trees for timber and wood pulp. The forest ecosystems are also important in other ways. They can influence the climate: by holding moisture like a sponge and releasing it slowly and steadily, they can prevent floods and droughts. They can even increase the water supply by catching moisture from the clouds.

Yet, despite their enormous value, we are bulldozing and burning the tropical forests to provide more land for farming and ranching and to build roads and towns. They are disappearing so fast that over half the world's tropical rain forests have been lost already. Some scientists estimate that we are destroying them at the rate of 40 hectares (100 acres) every minute. If that rate continues, they will be gone completely by the end of this century.

If the forests are allowed to disappear, so will hundreds of thousands of animal and plant species. They will become extinct and we will lose many invaluable sources of food, medicines and raw materials for industry. Preventing this from happening is one of the most urgent and important conservation problems facing the world today.

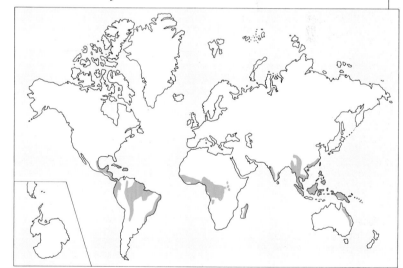

The world's rain forests
These forests occur in a broken strip stretching from South America, across Africa, to South-East Asia. Although they cover only a small percentage of the world's land mass, their richness in wildlife is unique.

Jungle cats

Hunting for the fur trade is pushing many of the rare, beautiful and little-understood creatures in the cat family to the brink of extinction. When one species becomes too scarce for hunting to be commercially viable another, more common one is exploited; and so the process continues until, species by species, they are hunted almost to extinction.

Tiger

Fifty years ago there were about 100,000 tigers living in Asia. Wherever there was water, prey and somewhere to hide, there were tigers. But they had a reputation for being dangerous, reportedly killing as many as 1,000 people a year in India alone, and so they were relentlessly hunted, trapped and poisoned. People who hunted tigers were respected for their bravery (even though with a modern rifle they could remain several hundred metres away) and the

tiger skins began to fetch large sums of money. Tiger hunting became a popular sport. So many were killed that several Indian maharajahs, in the first half of this century, claimed to have shot over 1,000 tigers each.

By 1970, there were 4,000 tigers left. Not only had tens of thousands been killed by hunting, but the population was also under pressure from habitat destruction. They had already disappeared from the island of Bali and there seemed little hope of saving the species from total extinction. Then came Operation Tiger. This massive effort became one of the greatest conservation success stories of all time. The species was protected by law and special tiger reserves were established right across its range, in India, Bangladesh, Nepal, Bhutan, Malaysia, Thailand, Indonesia, USSR, China and Korea. As a result there may now be as many as 7,500 tigers in the wild, and the numbers are still rising.

Man-eating is still a problem in some areas, but usually tigers prefer to give people a wide berth. Man-eaters are normally sick or wounded animals that cannot hunt their usual prey. Nowadays there are about 40 deaths a year in India.

Leopard cat

Smaller than the tiger, the leopard cat is about the same size as its domestic relatives, but moves so gracefully and elegantly that the average house cat looks almost clumsy in comparison. Superbly adapted to jungle life, they are excellent climbers and can even move along thin branches without difficulty. They are also good swimmers and have managed to reach and populate many offshore islands around their main homes in Sumatra, Borneo, Java, Taiwan, Japan and the Philippines.

Although sometimes seen during the day, leopard cats are most active at night and at twilight, when they go hunting. They spend most of their time in trees, in search of birds and reptiles, but they are opportunistic feeders and also hunt on the ground for hares, rodents and young deer.

Like most small cat species, leopard cats are hunted by man. Some populations have been considerably reduced in recent years. Their beautifully marked coats are much in demand by the fur trade, and since the intricate pattern-matching procedure requires enormous numbers of skins for each garment, they are under intense pressure.

Leopard cat (*Felis bengalensis*) *above*

ENDANGERED

Indian tiger (*Panthera tigris*) *right*

Jaguar

This attractive South American cat has disappeared from much of its former range due to habitat loss and hunting for its spotted coat. The height of demand for jaguar skins came in the 1960s, when 15,000 of the animals were being killed each year in the Amazonian region of Brazil alone. Jaguars are normally solitary animals, only meeting others of their kind to breed. Two or three cubs are born in a litter and they grow very quickly, attaining full adult size at about three years old.

ENDANGERED

Jaguar (*Panthera onca*)

Deadly hunters

Arrow-poison frogs are tiny, colourful animals that live in the understorey of Central and South American jungles. Their attractive appearance is deceptive, however, because in their bodies they have a poison so strong that a tiny drop is enough instantly to paralyse, or even kill, an animal as large as a monkey. The brilliant colours are there to advertise to predators that these frogs are highly poisonous and should be disregarded as potential meals. The fact that arrow-poison frogs are so common in some areas is a good indication of how effective this defence mechanism can be.

The name 'arrow-poison frog' is derived from the use of the frog's poison by South American Indians. The animals are impaled on a stick and held over a flame until the poison oozes out of their skin. It is then collected in a container, allowed to ferment and, as the name suggests, used on the tips of arrows for hunting. The poison, called batrachotoxin, acts on the heart and nervous system to immobilize hunted birds and monkeys.

It is interesting that the most reliable way of distinguishing the various species of arrow-poison frogs is by analyzing their poison. Individual species are often difficult to identify by sight and can vary considerably in size, colour and behaviour.

Arrow-poison frog

A skilful mimic

Katydids are fairly common nocturnal insects found in parts of Europe and the United States. The males produce shrill chirping sounds by rubbing a 'scraper' on the base of one forewing against a 'file' on the other. Many katydids expertly mimic leaves, with their flattened bodies and accentuated veins on the forewings (to simulate the veining of a leaf). They can jump only short distances.

Spike-headed katydid
(*Copiphora* sp.)

GREEN PYTHON

DATA

SPECIES
Green python
(*Chondropython viridis*)

CLASSIFICATION
Reptile (snake)

DISTRIBUTION
New Guinea, the Solomons, the
Aru Islands and northern
Australia

HABITAT
Rain forests

SIZE
Approximately 2 metres (7ft)

FOOD
Birds and tree-living mammals

The green python is one of the most sought-after snakes in the world. Its spectacular colouring is popular with collectors and it is traditionally eaten at the climax of every wedding feast in Papua New Guinea.

This python belongs to a group of primitive snakes, including all the pythons and boas, which kill their prey by constriction (squeezing). The green python feeds on tree-dwelling animals and has greatly elongated teeth that can gain and maintain a firm hold on its prey. Although its bite is often used in self-defence, and is important in the initial attack, it is not poisonous.

This snake can detect prey in complete darkness by smell and with the aid of its own directional heat-sensors. Its tongue leads small scent particles in the air to a very sensitive smelling organ in the roof of its mouth, while devices on both lips enable it to detect differences in heat. It could easily detect a human hand held a metre or more away, simply by the warmth given off.

Green pythons have fairly good vision but can only distinguish outlines well at short distances. They therefore cannot tell the difference between an animal standing completely still and any other lifeless object simply by sight. However, as soon as the animal moves, the snake reacts remarkably quickly. It strikes with its mouth open and immediately throws coils around its victim to crush it, killing by preventing the prey from breathing. The snake swallows its victim whole (by dislocating its jaws to open its mouth very wide) and digests it during the course of several days. The python may then go for many months without eating, since it requires little more than its own weight in food each year.

The green python lives in New Guinea, the Solomon and Aru islands and northern Australia. There is another, almost identical, snake called the emerald tree boa (*Corallus caninus*), which lives in northern South America. Both species are brilliant green in colour, with whitish or yellowish bands for camouflage; both have prehensile tails and are perfectly adapted to life in trees; and both have similar behaviour patterns. The only difference is that, as with all boas, the South American snake produces live young, whereas the python lays eggs. This is an extraordinary example of parallel evolution, where two species are almost identical even though they live on opposite sides of the world.

PARADISE FLYING SNAKE

Many different kinds of snake live high up in the rain-forest trees and rarely, if ever, come down to the ground. Most of these are forced to move around their difficult terrain very slowly, carefully picking their way through the tangle of branches and leaves.

The paradise flying snake (*Chrysopelea pelias*), of Borneo, is an exception. Along with other members of its exceptionally handsome group, it is able to 'fly' or glide with great speed from branch to branch, travelling distances of fifty metres or more in a few seconds. This is by no means true flight, in which altitude is maintained or even gained, but it is an incredibly effective mode of transport. To some degree, at least, the snakes can determine where they land by steering themselves with swimming movements in the air. They flatten their bodies, until they look more like ribbons, and draw themselves into S-shaped curves. This considerably

reduces the velocity of the fall and gives an effect which is very similar to parachuting.

Flying snakes are tremendously skilful climbers and are also able to move around at great speed within the trees. Sharp ridges on their bellies help with climbing and are used to gain 'footholds' on the relatively smooth bark of the coconut palms and other trees.

Practically every such palm tree has its own resident lizard population (chiefly made up of geckos and skinks), which forms the main source of food for paradise flying snakes. The snakes will often remain in the same tree for months or years without having to move farther afield. It is possible to find a series of moulted skins, all from the same snake, in a single tree. It is therefore likely that their 'flying' capabilities are used most often to escape the clutches of hawks and eagles, rather than to speed through the jungle canopy in pursuit of prey.

DATA
SPECIES
Paradise flying snake (*Chrysopelea pelias*)
CLASSIFICATION
Squamata (snakes and lizards)
DISTRIBUTION
Borneo
HABITAT
Jungles, particularly coconut palms
SIZE
1.5-2 metres
FOOD
Lizards, especially geckos and tree skinks

BIRD-EATING SPIDER

DATA

SPECIES
Bird-eating spider (sp.)

CLASSIFICATION
Spider (arachnid)

DISTRIBUTION
Most warm and hot countries

HABITAT
Wide range, including rain
forests

SIZE
Body length up to 9cm (4in),
total length up to 25cm (10in)
including legs

FOOD
Insects and small reptiles, birds
and mammals

Many people have an irrational fear of spiders. The reasons for their fear are unfounded because very few species can actually be considered dangerous. All 40,000 known species of spiders, with the exception of a single small family, possess 'venom' glands, but the venom itself is only effective against their normal small prey (usually insects) or is used only for digestion.

Bird-eating spiders, of which there are about 600 species, are the largest of all spiders, with bodies nine centimetres (nearly four inches) or more from end to end. Including its legs, a spider 25 centimetres (10 inches) long is not uncommon. Even so, only a few species have a lethal bite and the majority are not even capable of puncturing human skin.

Collectively known as tarantulas by many people, bird-eating spiders live in the southwestern United States, Central and South America, many Caribbean islands, large areas of Africa, Australia and India, and on Madagascar, New Guinea and Sri Lanka. In fact, they are found in most warm and hot countries around the world. They spend most of their time under stones, bark and leaves on the forest floor, coming out to hunt at night. Whereas other spiders feed mostly on insects, their size allows them to tackle small reptiles, mammals and birds as well. They do not build webs, which many other spider species use to trap food; instead they actively search or lie in wait for potential prey.

When attacked themselves, bird-eating spiders rise on to their four hind legs and bare their fangs at the intruder. Normally, though, they are calm and docile animals and will rarely attack or bite anything other than potential food.

These large, thickly-haired spiders are the most primitive of all species in their diverse group. They are very long-lived animals – while the males survive only about four years, or even less, the females can sometimes survive for more than 20 years.

ATLAS MOTH

With a wingspan of up to 25 centimetres (10 inches) and a rapid and agile flight, the spectacular atlas moth is often mistaken for a bird. Found in the rain forests of South-East Asia, this giant tropical moth is about as large as an insect can grow. The insect breathing system, which relies on a series of air tubes running from every part of the body to holes along the sides, considerably limits maximum attainable body size. The system works excellently over short distances, but as the length of the tubes increases the less efficient it becomes, until the body is so big that it doesn't work at all. This is why no butterflies or moths can grow as large as, for example, an eagle.

The male atlas moths have huge feathery antennae – the largest of any butterfly or moth – which they use for touching, assessing wind speed, gauging temperature and humidity and, most important of all, detecting the females' distinctive scent. The feathery structure increases the surface area available for scent reception, enabling the male insect to get a positive 'fix' on the location of the female, even if she is as far as two kilometres (one and a quarter miles) away. The antennae can respond to minute dilutions which are quite undetectable to the human nose, down to only one molecule of scent per cubic metre of air.

To illustrate the importance of scent to atlas moths, an empty box which a few days earlier had contained a female was placed in a suitable location in a forest. It was quickly sought out by the males – but they ignored a similar box in which females were kept under thick glass covers. Although the females were clearly visible, their scent could not be detected.

Female atlas moths have much smaller and unbranched antennae, since they do not have to locate the males. They are more sluggish and, while the males are often active by day, females fly only at night and then only for the purpose of laying eggs.

The caterpillars are also giants, many centimetres long. They pupate between the leaves of their food plants in spindle-shaped cocoons, fastened to the twigs with strong threads for security. The cocoons are so large that they look more like the fruit of the tree than a structure made by an insect.

The adult moths, when they emerge from their cocoons, do not live for very long. In both sexes, the mouthparts are completely non-functional,

DATA	
SPECIES	Atlas moth (*Attacus atlas*)
CLASSIFICATION	Insect (lepidopteran)
DISTRIBUTION	South-East Asia
HABITAT	Rain forests
SIZE	Wingspan up to 25cm (10in)
FOOD	Adults do not eat; caterpillars eat leaves of fruit trees

which means that they are unable to eat. The adults therefore have to live on food reserves stored up during the caterpillar stage of their life cycle. The male moths die soon after they have mated and the females after they have laid their eggs – which could be a matter of a few days or, in some cases, only a few hours.

BIRDWING BUTTERFLY

Butterflies are among the most colourful and most noticeable of all tropical rain forest inhabitants. Even aircraft pilots flying over the jungles of South America have noticed the brilliantly-coloured wings of some species. Among the largest and most beautiful of all are the birdwings, found only in South-East Asia. All 12 species are rare and, unfortunately, much sought after by collectors; they cost a great deal to buy since they are difficult to catch, spending most of their time in the tops of giant jungle trees. The largest butterfly in the world is the Queen Alexandra birdwing of New Guinea, which can have a wingspan of over 28 centimetres (11 inches).

MOUNTAIN GORILLA

BLACK GIBBON

A group of excited gibbons can be heard calling several kilometres away, their long and distinct sounds are so loud. Usually they call or 'sing' when other groups venture too close, or they may display by breaking branches and performing spectacular acrobatics. Their scientific name, *Hylobates*, actually means 'dweller in the trees' and they exceed all other animals in their agility.

Despite their impressive chest-beating displays and sheer size, gorillas are normally quiet and docile animals and not nearly as dangerous as they are made out to be. They can fight ferociously if they feel they really need to, particularly if disturbed by the presence of other gorillas or humans, but prefer to threaten rather than attack.

There are three sub-species of gorilla, all living in the dense, moist rain forests of central Africa. The rarest of the three is the mountain gorilla, of which only about 365 survive in the wild. There are more individuals of the eastern lowland and western lowland sub-species – their numbers are counted in thousands – but they are by no means common. Poaching, forestry and the encroachment of agriculture into their forest homes have all taken their toll, and gorilla numbers have decreased drastically in recent years. Indeed mountain gorillas, which have longer and silkier fur than the lowland varieties, are among the rarest animals in the world.

Mountain gorillas live only in the Virunga Volcanoes region straddling the borders of Zaire, Rwanda and Uganda in central Africa. Troops of about 15 individuals are most common, each with a dominant male or 'silverback' (so-called because of the silvery-white saddle of hairs which develops on his back), several other males, females and young. Together, they roam over distances of up to a kilometre or more each day. The adults spend most of their time feeding and grooming, but young gorillas in the troop play much like human children, constantly climbing trees, swinging on branches, chasing each other and imitating their parents.

Most of a gorilla's time is spent on the ground, walking on all fours or, very occasionally, upright on its legs like a human. However, gorillas are quite capable of climbing and will readily do so to collect food or to view the surrounding terrain. They are almost entirely vegetarian, choosing mainly shoots and leaves, though they will also take nuts, berries, fruits and occasionally grubs. In captivity they will eat meat. Their fingers are thick and stubby, so they often use their lips or teeth when removing skins and pips.

Feeding times, movements and other activities are all controlled by the dominant silverback and his instructions are co-ordinated by grunts and barks from other members of the group. As soon as it begins to get dark, he leads the troop in the evening ritual of nest-building. Each individual spends about five minutes building a new nest to sleep in every night, either on the ground or in the trees.

Silverback mountain gorilla *left* **Female mountain gorilla with young** *below*

TARSIER

As soon as the sun has set, and the island forests of Indonesia and the Philippines are in darkness, a tiny and attractive jungle animal comes to life. No bigger than a human hand, the tarsier is

perfectly adapted to its nocturnal life in the trees. Its eyes are larger than those of any other animal of the same size, to help it see in the dark. Its ears are extraordinarily sensitive, constantly moving and ready to detect even the faintest noises in the darkened trees.

Tarsiers feed mostly on insects but they will also eat lizards, bats and even scorpions or poisonous snakes. Individuals have their favourite food and will always eat some items in preference to others. As soon as they hear a promising sound they turn their heads (often almost far enough around to see directly behind themselves) and focus on the unsuspecting victim in the dark. Then they jump on it. The name 'tarsier' actually means elongated tarsal, or ankle, and the jump is when the enormous frog-like limbs come into their own. With a movement much like that of a tree frog, they spring from their perch (often backwards), twisting their bodies and tucking in their arms and legs, until they land feet first on the next trunk or branch, often over a metre away. To hang on, they use their slender, sucker-tipped fingers and toes which enable them to grip strongly on almost any surface.

The victim is usually pinned down by one or both hands, but tarsiers will often use their fingers like a butterfly net if the prey is airborne. They kill, or at least immobilize, the animal with several bites from their large, needle-sharp teeth, which are also sharp enough to frighten off the tarsier's own enemies. The victim, often caught on or near the ground, is carried in the mouth to a perch to be eaten.

Young tarsiers can be born at any time of year, though births are thought to be more common at the end of the rainy season, between February and April. They are born fully formed with open eyes and the adult's characteristic velvety fur. (The three different tarsier species look very similar but can be distinguished by the tuft of hairs on the ends of their tails.) The single young quickly learns to cling to its mother's belly (though she sometimes carries it around in her mouth) and spends its daytime sleeping with her in a safe clump of dense vegetation. Before hunting, the mother will 'park' her offspring on a nearby branch and keep in touch with it by clicking and whistling calls. The young animal can climb around when less than a day old and, within a very short time, is leaping about in the trees like its acrobatic parents.

DATA
ENDANGERED
SPECIES
Tarsier (*Tarsius* sp.)
CLASSIFICATION
Mammal (primate)
DISTRIBUTION
Indonesia and the Philippines
HABITAT
Dense rain forests and mangroves
SIZE
Body length 8.5 to 15cm (3½ to 6in) plus tail of 20 to 26cm (8 to 10½in)
FOOD
Mainly insects, but also lizards, bats and even scorpions and poisonous snakes

and watching a terrible battle during the Spanish conquest of South America. After the fighting had finished, they flew down to look after the injured and dying Indians but, while they helped, their breast feathers were stained red with blood. The birds are said to have kept the beautiful combination of emerald green and crimson feathers ever since this incident.

Quetzals were once sought after by the Aztecs and the Mayas who, like many other peoples throughout the world, prized brightly-coloured feathers as items of personal adornment. No ordinary person was allowed to possess them – only the highest dignitaries could wear them, as plumes in their ceremonial dress. They captured the birds alive, plucked their beautiful long tail feathers, and released them. Since the birds normally shed and regrow these feathers after each breeding season, they were unharmed. The quetzal was even worshipped by these people as the god of air. Their deity Quetzalcoatl, the god of civilization, was symbolized as a serpent with quetzal feathers.

Quetzals, also called 'resplendent trogons', are the most famous, and among the rarest, of the trogons – a group of about 35 different species found in both New and Old World rain forests. Several trogon species are closely related to the quetzal, such as the contrasting pavonine quetzal, whose upper tail covert feathers are no longer than the actual tail. As a group trogons are normally solitary birds, eating fruit, berries and insects (which they often find by following troops of monkeys and using them as 'beaters'). Occasionally they take frogs and lizards. Quetzals are unusual among trogons in preferring to associate in small groups. They particularly favour the fruit of *Ocotea*, a member of the laurel family, which they swallow whole, regurgitating the large stone later.

Most trogons populate the lower levels of the canopy, nesting in tree cavities and often using the same holes as ants or wasps in which they lay their two or three eggs. The quetzal's eggs are light blue and are always laid in pairs in decaying tree trunks, in holes which greatly resemble those of woodpeckers. There is only one entrance and the male's tail feathers often get very badly damaged as he clambers in and out. The clutch is incubated at night and midday by the female, and during the morning and afternoon by the male, for a total of 17 or 18 days. The young, born naked and with their eyes closed, remain in the nest for a full month and leave it only when they are fully fledged.

DATA

ENDANGERED	
SPECIES	Quetzal (*Pharomachrus mocino*)
CLASSIFICATION	Bird (trogon)
DISTRIBUTION	Central America
HABITAT	Mountain rain forest
SIZE	Up to 1.5 metres (5ft), inc. tail
FOOD	Insects, fruit, berries, sometimes small frogs, lizards and snails

QUETZAL

In the mountain jungles of Central America lives one of the most colourful of all tropical birds, the quetzal. Its plumage is so extraordinary that its very existence was doubted for many years. From southern Mexico to western Panama its spectacular emerald and crimson coloration is a well-known, albeit rare, sight. It has even become the national emblem of Guatemala, and gives its name to the country's unit of currency.

There is a Guatemalan Indian legend which says that, many years ago, quetzals were entirely green in colour. According to the legend, one day a huge flock of the birds was hiding in a tree

BIRD OF PARADISE

In the densely forested areas of Papua New Guinea and northern Australia there lives a group of birds with a bewildering variety of dazzling plumages. So fantastic are their colours that the birds' discoverers believed them to be from Paradise, because nothing else on earth resembled such brilliance.

In fact, birds of paradise, as they were aptly called, are relatives of the more down-to-earth crows and starlings. Only the males have the famous striking plumage; the females are rather dull. Males undertake elaborate displays which enhance the visibility of whichever part of the plumage is the most colourful, with a spectacular overall effect designed to attract the female for mating. Several species, such as the king bird, even display by hanging upside-down underneath a branch, enabling them to show off the bright undersides of their wings and patches of breast feathers. Male birds often become so obsessed with displaying their costumes that they do little else, even after mating. Each male's aim is to attract as many females as he can before the end of the season.

Over 40 different species of birds of paradise have been identified, all very closely related. Their feeding habits are unspecialized; they eat mainly fruits, insects and spiders high up in the jungle canopy. The males of different species are easily distinguished from one another, but the females are remarkably similar. Breeding between different species, normally a very rare occurrence in the animal kingdom, is surprisingly frequent among these birds and it is not unusual to see several species displaying communally. They display in particularly prominent trees, which have probably been used for the purpose for decades or even centuries. A flock of a dozen birds or more gathers, shrieking and displaying, each awaiting his chance to perform on the dancing branch.

The males take no part in nest-making, incubating the eggs or rearing the chicks. The female builds a well-constructed nest of leaves about four to ten metres (13 to 33 feet) above the ground in dense thicket. She incubates her two to four eggs for roughly three weeks. By the time the young chicks leave the nest the tail plumes of the males will have moulted, but these regrow in time for more courting in the next season. The natives of New Guinea have long used the feathers for their headgear, particularly the blue breast feathers of blue birds and the golden rain of the greater bird of paradise. The birds' populations can withstand such minimal exploitation but several species came close to extinction when their feathers were in great demand for commercial hat-making. Although the export of these feathers is now mostly banned, poaching continues in some areas; and with the general destruction of their rain-forest homes, several species of birds of paradise are now seriously threatened.

DATA	
SPECIES	Bird of paradise (sp.)
CLASSIFICATION	Bird (passerine or perching bird)
DISTRIBUTION	New Guinea and neighbouring islands, north-eastern Australia
HABITAT	Rain forest
SIZE	17 to 120cm (7 to 48in) depending on the species
FOOD	Berries, fruit, insects and spiders

Lesser bird of paradise
(*Paradisaea minor*) far left

TOUCAN

Toucans are well known for their large, often colourful beaks. In some species they can be up to 23 centimetres (nine inches) long, exceeding the length of the body. Although not too heavy, the beaks are extremely strong and, apart from being long enough to enable the bird to pluck distant fruit with ease, are used to intimidate potential predators. There are about 40 species of toucan, all found in Central and South America, where they inhabit montane and lowland rain forests and scrubland.

TEMPERATE WOODLANDS

Temperate woodlands have probably influenced man's progress more than any other habitat in the northern hemisphere. The trees produce strong and beautiful timber and harbour a wealth of animals and plants, while the soils in which they stand are rich and ideal for planting crops. These characteristics, however, have been the cause of their downfall. Only remnants of the world's original temperate woodlands survive today because people have cleared them to provide farmland and timber. Most of the remaining forests have been profoundly changed by the establishment of permanent clearings to provide space for human settlements and yet more agricultural land.

The decline of temperate woodlands

In Britain, for example, temperate woodlands once covered 70 per cent of the total land area. Today they cover less than 10 per cent. Vast tracts of oak, beech, elm, birch and ash that once harboured 5,000 or more different species of animals and plants, and that have been in existence since the end of the last Ice Age, have been unable to survive the pressures of the 20th century. Britain has lost thousands of square kilometres of these ancient woodlands since the Second World War alone and today the remaining fragments are disappearing faster than ever.

It is a similar story over most of these forests' range, which extends across central and western Europe eastwards through Russia, in a belt between the northern coniferous forests and the southern steppes; in many parts of North America, central China and Japan; and, in the southern hemisphere, in parts of South America, extreme southern Africa, south-west and south-east Australia and New Zealand.

Despite their wide distribution and the wide range of climatic conditions in which they live, temperate woodlands have many common characteristics. Structurally, they are very similar. The trees are generally smaller than those in tropical rain forests, though large species such as oak and beech form the canopy and dominate the community. They block out much of the sun and have a shading effect, so it is mostly shade-tolerant trees that live beneath. Maples, birches and similar species form the secondary layer; beneath these are dogwoods, hawthorns, hollies and other shrub-like plants; these in turn shade a great diversity of flowers and plants on the woodland floor.

This range of trees – over a thousand species worldwide – and other plants, all living within a single wood, provide many different habitats. Most woodland animals are adapted to living in just one of these habitats. In the rides and glades are creatures that like grassland with plenty of cover, while other species choose one of the other woodland areas – along the woodland edge, right in the centre of the wood itself, in the tops of the trees or under the roots. A single oak tree can support well over 300 different insect species alone, compared with fewer than a dozen on a conifer. Each insect is highly specialized and only capable of living on a leaf, twig, root or other particular part of its host, but together they form an incredible web of interdependent species all relying on one another for their survival.

English woodland *above*
Bluebells and sycamore seedlings cover the floor.

Beech woods *opposite*
Ancient beech trees in the New Forest, southern England.

The woodland in winter

Another characteristic of temperate woodlands is that they are comprised of mostly 'deciduous' trees – species that drop their leaves during the winter period. Severe cold would damage the leaves and, if there are strong winds, the resistance caused by the foliage would break branches or even fell whole trees. Instead, a deciduous tree stores next year's leaves and flowers in well-protected, waterproof buds. Deciduous species are sometimes called 'broad-leaved' because of the broad, flat leaf shape which contrasts with the needle-like or scaly leaves of a coniferous tree.

There is little sign of life in the temperate wood during winter. Some animals, such as hedgehogs and wood-chucks, are hibernating; others remain warm and secure in underground burrows or holes in the trees, emerging only occasionally to hunt and forage for food. The trees themselves are mere skeletons and the once green and lush woodland floor is carpeted with dead brown leaves. These will be broken down by unobtrusive beetles, earthworms and other invertebrates, plus fungi and bacteria, which act as recycling agents as they constantly return the valuable nutrients and other components of the dead forest litter to the soil for living plants to re-use.

As the winter's grip on the woodland begins to relax, sunny days in early spring bring flowers such as bluebells and primroses into bloom and trees bud into leaf. Birds begin to sing and the woodland comes alive every morning with the dawn chorus. The bats begin to emerge from their roosts to fly among the trees in search of early moths, while squirrels scamper through the now-leafy branches.

By summer the woods are filled with animals and their young. The vast majority of the new-born individuals will succumb to hunger or predation before they mature into adults, but the remainder will join their elders in gathering food in readiness for winter. Then, as the summer fades away, the trees burst into colour and lose their leaves, days shorten and the temperatures fall, and the annual cycle of the temperate woods begins again.

The world's temperate woods *left*
Areas of temperate woodland can be found in central and western Europe, the eastern USA, China and Japan, and in many other parts of the globe.

Woodland beetles

There are over 300,000 different species of beetle in the world, many of which inhabit temperate woodlands. Two of the most interesting are the familiar ladybird and the large stag beetle.

Seven-spot ladybird

Ladybirds are among the most useful insects in the world to gardeners. Both the adults and their larvae feed almost entirely on aphids, such as greenfly and blackfly, and other small bugs. Ladybirds have even been raised on special 'ladybird farms' for sale to horticulturalists, to help them fight these tiny but major garden pests. Many millions of ladybirds have been released into the orange groves of California and other fruit-growing regions for the same reason.

Ladybirds are brightly-coloured beetles with very shiny, domed bodies. There are over 4,000 species and they are found in every inhabited continent of the world. As with most other members of the beetle group, they have tough outer wing cases which have developed from the front pair of wings to form a protective shield over the hind pair. These front wing cases are raised clear of the hind wings in flight – they can easily be seen as the animal takes off. The cases are extremely variable in both colour and markings, though they do tend to follow a general pattern – red with black spots, black with red spots, yellow with black spots, and so on.

There is a good reason for this bright coloration. The bold pattern is there specifically to warn birds and other potential predators that the ladybirds taste horrible and should be left alone. When attacked and seized, a ladybird produces a red, bitter-tasting liquid which not only tastes horrible but also stains and smells for some time afterwards. Young, inexperienced birds will normally catch and attempt to eat a ladybird or two but they soon learn to leave them alone.

Ladybirds lay small, yellow eggs in groups on the undersides of leaves which are infested with promising colonies of aphids. The slate-blue larvae hatch out to find their living meals ready to eat; they spend the short three weeks of their larval lives devouring the hapless pests. Rather than randomly searching the leaves, larvae concentrate their activities on the leaf veins, where the aphids tend to cluster. In this way a single larva may eat 50 aphids in a day.

Stag beetle

The world's beetles range in size from 0.1 to 15 centimetres (¹/₂₅ of an inch to six inches). The largest of all is the powerfully-built Goliath beetle (*Goliathus giganteus*) which is no less than 8 million times heavier than its smallest relative.

Stag beetles are the largest beetles in Europe. They tend to vary in size but many specimens reach over five centimetres (two inches) in length. Their name is derived from the large antler-like mouthparts, or mandibles, of the male. If disturbed, a male will rear up and open these antlers wide to frighten off a potential enemy. It is quite a fearsome sight but the gesture is largely bluff because the muscles of these giant structures are far too weak to cause a painful bite. They are just strong enough to hold on to the female during mating, though they are also used in fights between rival males.

In contrast, the mandibles of the female are much smaller and can give a nasty nip. Indeed, the two sexes look so dissimilar that they were once believed to be different species.

Since the male's mouthparts are so unwieldy and impractical they cannot be used for chewing, and in fact both sexes are unable to eat anything. At the most, they may feed on a few plant fluids. The fat, white larvae, on the other hand, spend most of their time eating. These young grubs have tough jaws specially designed for chewing dead and decaying wood.

Seven-spot ladybird
(*Coccinella septem-punctata*)
below
Stag beetles (*Lucanus cervus*)
bottom

An introduced species

Fallow deer have been introduced by man to woodlands in many areas of the world, and they are now found throughout western Europe and in parts of Africa, Australasia and North and South America. The coat colour varies enormously from animal to animal and ranges from nearly white to nearly black. The most common coloration is a rich fawn with many prominent white spots in summer and a duller grey-brown coat with barely-detectable spots in winter.

Fallow deer buck (*Dama dama*) *left*

Tree-top butterfly

Purple emperor butterflies are rarely seen because they spend most of their time at the tops of oak and ash trees. They return to their favourite spots on particular leaves or branches again and again, sometimes chasing away other butterflies and even birds which venture too near. Unlike most other butterflies purple emperors do not seem to be attracted to flowers but prefer to feed on decaying animals and cow-pats. They also seem to be fascinated by cars and will often attempt to get into a parked vehicle or rest on the bonnet.

Male purple emperor (*Apatura iris*) *far left*

Self-defence methods

Grass snakes are not poisonous and very rarely bite, but they do have elaborate methods of self-defence. Occasionally they will strike at an attacker with the mouth closed, but more often they hiss loudly and release the evil-smelling contents of their aptly-named 'stink glands' situated at the anus. If they are still being threatened they may feign death by lying on their backs with their mouths wide open and their tongues hanging out. In this state the snakes will allow themselves to be handled, without putting up any resistance, for half an hour or more; but within 30 seconds of the attacker moving off they quickly come to and disappear into the undergrowth.

Grass snake (*Natrix natrix*) shamming dead

25

WILD BOAR

DATA
SPECIES Wild boar (*Sus scrofa*)
CLASSIFICATION Mammal (artiodactyl – hoofed mammal)
DISTRIBUTION Many parts of Europe, through to south-eastern Siberia, and southern Asia
HABITAT Variety of habitats, wherever there is dense vegetation for cover; especially in temperate woodlands
SIZE Head and body length 1.5 metres (5ft), plus tail of up to 40cm (16in)
FOOD Almost anything including roots, bulbs, fungi, nuts, snails, mice and young rabbits

Wild pigs roam forests, meadows and swamps in many parts of the world. They are the ancestors of domestic pigs and have remained virtually unchanged since they first appeared in prehistoric times, over 45 million years ago.

The wild boar is the only wild member of the pig family found in Europe. It became extinct in Britain in the 17th century but is still found in many countries of mainland Europe and has been introduced by man to such places as southern Scandinavia, where it had never occurred before.

Wild boars are most active at dusk and dawn or during the night, when they can be heard contentedly grunting as they forage. They can also be located by their distinctive smell – rather like damp oak leaves – which is sufficiently strong for the human nose to detect quite easily. The animal eats almost anything and has become a considerable pest in some areas, raiding crops such as grain, potatoes, melons and grapes. It adores wild garlic but is not fussy about its diet and will take fungi, roots, bulbs, green vegetation, nuts, snails, mice, carrion and anything else available. For some reason, one of the few things it will not eat is wild daffodils.

The males are generally solitary animals, except during winter when they are seeking out females for mating. This is when they use their strong tusks in fights over potential mates. As soon as a pair has mated the male departs, leaving the female to cope with rearing the family. Usually there are four to eight piglets in a litter, born in a lair under a tumble of rocks or a fallen tree. Once the piglets leave the lair they often chase the mother and butt against her side until she allows them to drink her milk. A sow is often accompanied not only by these young animals but also by her nearly full-grown young of the year before. Together these family groups form small herds known as 'sounders'. The young are fiercely defended from attackers by their mother and they follow her for up to two years before searching for a mate of their own. The young males generally leave first and wander alone for several years until they are big enough to compete successfully with adults.

As with other members of the pig family, wild boars are swift runners and good, strong swimmers. They love wallowing in mud pools, both to cool off and to rid themselves of insects, and will do so for many hours if the opportunity arises.

KOALA

Although the koala looks like a bear, and is often described as one, it is not a bear at all. It is a marsupial, or pouched mammal, more closely related to kangaroos, wombats and bandicoots than to the true bears such as grizzlies and polar bears.

Once common in many parts of Australia, large numbers of koalas have been killed over the years for food and for their thick, warm fur. In 1924, when the hunting reached its peak, no fewer than two million skins were exported from Australia – and many more koalas were losing their homes because of forest fires and clearance of their forest homes to make way for farmland. Today, though, they are protected by strict laws and are slowly increasing in numbers in many parts of their range.

Koalas live almost entirely in eucalypts, or 'gum' trees. In fact, they spend so much time in them and eat so many of their leaves that they actually smell of eucalyptus. A koala spends as much as 20 hours each day dozing or sleeping, usually curled up in a tree fork high above the ground. It then eats for several hours during the night, consuming as much as a kilogram (over two pounds) of leaves and bark. They are very fussy feeders: with about 350 different species of eucalypts to choose from, they will eat the leaves of only 20. Koalas are also very careful and cautious feeders. Sometimes, there seems to be no logical reason why they climb past several leafy branches before stopping to eat. But at certain times of the year eucalyptus leaves contain high concentrations of a poisonous acid and, although koalas can cope with small quantities of this potentially deadly chemical, they prefer to avoid it whenever possible.

These marsupials are excellent climbers, grasping the trunks and branches with their fingers and toes with the help of sharp claws which dig into the bark. They only occasionally come to the ground. When they do, it is merely to shuffle slowly to another tree which is too far away to climb or jump into.

Koalas are usually solitary animals but form small groups, consisting of one male and several females, during the breeding season. This is also the time of year when they are at their noisiest, making calls which sound like saws cutting wood in the forest. Normally they are silent; occasionally they make a pig-like grunt while foraging at night, or a loud wail when alarmed.

Baby koalas are born in spring or summer,

and at first are no bigger than a grape. Soon after birth the young animal climbs blindly into the mother's protective pouch on her belly, where it remains, feeding on her milk, for about six months. After this time it is strong enough to leave the pouch but still travels with its mother, on her back, until the following spring. By then young koalas are eating the eucalypt leaves and will soon be old enough to leave home and live alone. They begin to breed when about three or four years old.

DATA
SPECIES
Koala
(*Phascolarctos cinereus*)
CLASSIFICATION
Mammal (marsupial)
DISTRIBUTION
South-east Australia
HABITAT
Eucalyptus forests
SIZE
60 to 85cm (2 to 2¾ft)
FOOD
Leaves and young bark of eucalyptus, also mistletoe and box leaves

EUROPEAN BADGER

DATA

SPECIES
Badger (*Meles meles*)

CLASSIFICATION
Mammal (carnivore)

DISTRIBUTION
Widely distributed throughout Europe as far north as the Arctic Circle and east to Asia and Japan

HABITAT
Mostly temperate woodland but also scrub, hedgerows, quarries, sea cliffs, moorland, open fields and even in gardens and under buildings

SIZE
Up to 80cm (2½ft), plus tail of about 15cm (6in)

FOOD
Most important food is earthworms; also small animals such as mice, voles, young rabbits, slugs, snails, grubs, beetles; fruit, nuts, berries, grass and other vegetable matter

Badgers are fairly common throughout most parts of Europe and across Asia as far east as Japan. However, they are rarely seen because they normally only come out at night and spend the daytime hidden underground.

Badgers live in setts, which are built in a variety of places including woodland, hedgerows, quarries, open fields and even under buildings. Each sett consists of a complex labyrinth of tunnels and chambers lined with heaps of bedding consisting of dry vegetation. A sett has many different entrances, which are easily distinguished from a fox's earth by their large size, the heaps of earth mixed with bedding outside, and the well-marked paths leading to and from the sett.

A number of badger families often share the same home range and occupy the various setts within it. These can be quite crowded if several females all have cubs at the same time. A single litter of up to four cubs is born each year, usually between mid-January and mid-March. The young badgers remain below ground for about eight weeks but live and forage with the sow at least until autumn and often for the duration of the first winter.

The occupants usually emerge from their sett around dusk, though sometimes in the height of summer – and particularly in remote areas – badgers can be seen outside well before dark. They do not hibernate, but in winter their activity is considerably reduced and they may stay below ground for several days at a time. When they do appear, they take great care to survey their surroundings thoroughly before emerging. Badgers cannot see very well and use their eyes mainly for detecting movement, but their senses of smell and hearing are acute. When they are sure the coast is clear they move off in search of food, which is mostly earthworms but can be anything from beetles to berries.

Badgers have suffered considerably at the hands of man in recent years. In Britain some authorities believe that they are responsible for passing bovine tuberculosis on to cattle, and consequently many thousands of badgers have been killed in the worst-hit TB areas of the south-west. This has had no significant effect on the incidence of the disease in cattle and is now believed by many people to be unnecessary. Large numbers of badgers are also captured and slowly killed in an illegal 'sport' known as badger baiting. The badgers are handicapped in some way, for example by having a jaw broken, and then placed in a pit and set on by dogs to fight to the death. Despite these persecutions, however, badgers are holding their own in most parts of their range.

COMMON DORMOUSE

When the outside temperature falls and remains below about 15°C (60°F), which is usually some time in October, the common dormouse begins its long winter hibernation. Curled up in its nest on or under the ground, it sleeps until April and wakes only occasionally for a few days of activity. Hibernation is a dangerous time and many dormice die, either through starvation or predation by weasels, badgers, foxes, crows, magpies and others. Dormice are strictly nocturnal and spend most of their time in trees high above the ground searching for nuts, fruits and berries.

LEAST CHIPMUNK

There are 22 different species of chipmunk living in North America and one in Asia. They belong to a group of rodents which includes squirrels, marmots and prairie dogs and can be found in most open woodlands and city parks throughout their range.

The least chipmunk is the smallest of the group and the most widely distributed of the American species. Only the Siberian chipmunk (*Eutamias sibiricus*), which is found in Siberia, Mongolia, northern and central China and Korea, occurs over a wider area.

Chipmunks spend most of their time searching for food and sleeping. They spend the summer months gathering nuts, seeds, berries, fruit, grain and insects. Some of this food is eaten while the remainder is carried around in their cheek pouches, which act as shopping baskets. Some of the spare food is tucked under rocks and logs for emergencies, and the rest is stored in a specially-built larder under the ground, ready for winter.

As with all American species, the least chipmunk builds its winter nest in a chamber at the end of an underground tunnel. The tunnel itself measures about two and a half centimetres (one inch) wide and 1.5 metres (5 ft) long, and the chamber doubles as both a larder and a bedroom. Only the Asian species uses two separate rooms. During winter the animals, which have thicker, woollier coats to keep them warm, wake from time to time to feed from their store. Unlike ground squirrels (*Spermophilus* sp.) and many other hibernating animals, they do not live off accumulated body fat. The total length of time spent underground varies enormously, but the least chipmunk usually stays in its tunnel from October through to April.

Although they spend most of their time on the ground, all chipmunks can climb trees and bushes. Indeed, during the summer they rarely use their underground burrows, preferring to sleep at night in a nest in the branches of a tree, or even in a hollow log.

Females bear four to eight young in the spring. Both parents care for the offspring for about two months, after which the young chipmunks begin putting away their own food for the winter ahead.

DATA
SPECIES
Least chipmunk (*Eutamias minimus*)
CLASSIFICATION
Mammal (rodent)
DISTRIBUTION
North America, from Yukon to Ontario and Wisconsin, and mountainous areas of western United States
HABITAT
Variety of woodland, including temperate and coniferous
SIZE
Body length 9 to 11cm (3½ to 4½in), plus tail of 7.5 to 11cm (3 to 4½in)
FOOD
Fruits and seeds of various trees and herbs; occasionally mushrooms, insects, bulbs and birds' eggs

BLACK WOODPECKER

DATA
SPECIES
Black woodpecker (*Dryocopus martius*)
CLASSIFICATION
Bird (piciformes – woodpeckers and their relatives)
DISTRIBUTION
Many parts of Europe, across Siberia to Japan; parts of Asia
HABITAT
Deciduous forests
SIZE
45cm (18in)
FOOD
Ants, beetle larvae, wood wasps

At 45 centimetres (18 inches) from the tip of its beak to the end of its tail, the black woodpecker is by far the largest woodpecker in Europe. It is also the most distinctive. The male in particular is a very striking bird, being all black except for a bright-red crown and crest. It also has a far-carrying call and makes an extremely loud drumming sound as it pecks which can be heard several kilometres away.

The drumming is not connected with feeding. It is in fact a 'keep away' signal that a male makes to other males, proclaiming his ownership of the local territory. It is also a 'come hither' sign to any females within earshot. In most woodpeckers the drumming consists of ten or so blows delivered in rapid succession, often in less than a second, to a resonant object such as a hollow branch or even a metal plate on a telegraph pole.

Black woodpeckers live in a variety of woodland habitats but are particularly fond of old stands of conifers mixed with beech. Trees provide them with both nesting places and food. The female bird lays four or five eggs on the floor of a specially excavated cavity in a tree, lined only with a layer of wood chips. Incubation takes less than three weeks, after which the young are fed on a variety of insect adults and grubs (mostly ants and their larvae, and wood wasps) which their parents collect from the trees and on the ground. This sometimes involves boring huge holes in decaying wood during their search.

There are about 230 different species of woodpecker, found in most parts of the world. Although they vary greatly in both size and colour they have a great many characteristics in common. They are well adapted to climbing about in trees, with short legs and a stiff tail to act as support against tree trunks and other vertical surfaces. Their feet are designed for climbing and clinging, with two strong toes pointing forwards and two backwards, all armed with long, sharp claws. Even the bill is stout and chisel-shaped for drilling holes in trees.

The largest British representative of the group is the green woodpecker, at about 30 centimetres (12 inches) long. It does not drum as much as its relatives but it does make a loud, raucous laughing call which has given it the country name of 'yaffle'. At the other end of the size spectrum is the lesser spotted woodpecker, only just bigger than a sparrow, but which can drum at the rate of 15 or more blows each second. Between these two birds in size is the great spotted woodpecker, also a fast drummer, but louder than the lesser spotted.

The woodpecker's tongue is extremely long and sticky and designed to work like a spear. It has a hard, pointed tip, equipped with barbs, to hook and extract the grubs from their tunnels under the bark. The skull is specially shaped with a storage compartment to hold the tongue, and is also toughened to withstand the tremendous forces exerted on it when 'drumming'. It is so strong that scientists have been studying its structure in the hope that it might provide useful hints for improving the design of crash helmets worn by motorcyclists.

NIGHTINGALE

The nightingale is one of our most famous songbirds. Generations of poets have been inspired by its remarkably varied and loud song, which is delivered at night as well as during the day.

To the uninitiated, any bird that sings at night is a nightingale. However other thrushes, and also birds from other families such as nightjars, are nocturnal songsters. Once heard and distinguished, though, the nightingale's song is unforgettable. In the still of late evening it suddenly bursts forth with a combination of long, pure, flute-like 'peeuuu' notes, cracked and croaking noises and chugging 'juk . . . juk' sounds. Individual birds call to each other with a liquid 'wheet'; there are also the ticking 'tack . . . tack' notes common to other birds in the group, such as robins; and a harsh scolding call. Some birds are better singers than others, but in all members of the species the notes are loud, clear and endlessly variable. Unfortunately we are only treated to the nightingale concert from middle or late April to the end of June. Its Northern European counterpart, the nightingale thrush (*Luscinia luscinia*), is almost identical in appearance and its song is even louder but not so musical.

Nightingales live in temperate woodlands with thick undergrowth, but can also be found in overgrown hedgerows, shrubberies and large gardens. In Mediterranean Europe they often sing from open perches and are fairly easy to see, but in most other parts of their range they are inveterate skulkers and spend much of their time hidden in dense bushes. Apart from a quick flash of the conspicuously rufous tail as one darts out of hiding to pick up a beetle or a spider, these highly secretive birds are rarely seen. The easiest way to locate them is by their singing, especially after dusk or before dawn, when most other birds are silent. On a calm night, their rich sounds can be heard two kilometres (one and a quarter miles) away.

Nightingales are migratory birds and cover great distances flying between central Africa, where they spend the winter, and Europe, where they breed during the summer. The reason they migrate is because they feed almost entirely on insects, which are absent from the temperate woodlands during the cold season; unlike many other species, they cannot replace these with berries or other winter foods.

The males return to Europe first, usually some

time in early or mid-April; by the end of July many birds have already started their journey back to Africa, always flying at night. By the beginning of October almost all the stragglers have gone. In the space of those few summer months they have to establish territories, mate, build nests, rear their young and prepare for the arduous homeward journey.

The female builds the nest of dead leaves lined with grass, close to the ground and carefully concealed. She lays four or five olive-brown or olive-green eggs and incubates them alone for about two weeks. The male, however, helps in rearing the young, which usually hatch in late May or early June. The fledgling nightingales disperse even before they can fly, leaving the nest when they are only about 12 days old. It leaves them little time to prepare for their first journey back to Africa.

DATA
SPECIES
Nightingale (*Luscinia megarhynchos*)
CLASSIFICATION
Bird (passerine – perching bird)
DISTRIBUTION
From Britain to northern Africa, south-western Siberia and central Asia
HABITAT
Undergrowth of woods and thickets
SIZE
16cm (6¼in)
FOOD
Worms, spiders, caterpillars and a variety of other insects

CONIFEROUS FORESTS

In a broad band stretching across the northern parts of the northern hemisphere, and on the mountains and high plateaux of the south, are the coniferous forests. These are the tough and hardy forests of the world, able to survive long, severe winters and to grow in poor-quality soils.

Typical trees

There are two main types of coniferous forest, one found at high latitudes and the other at high altitudes. They both have similar characteristics with well-spaced, cone-bearing trees and a sparse undergrowth. Compared with tropical rain forests and temperate forests, there are comparatively few tree species in the coniferous zone. Depending on where the forest is, the dominant trees are generally pines, firs, larches, spruces, redwoods or hemlocks. In warmer areas, such as California and the Mediterranean, the cedars, cypresses and junipers tend to be more common. The southern hemisphere has no pines, firs, spruces, true cedars or cypresses but there are remnants of forests in which coniferous trees occur among other species; these include the yellow-wood of South Africa, the King William pine of Tasmania and the Chilean cedar.

Probably the most spectacular of all coniferous species are the redwoods in the Pacific coastal regions of British Columbia and the USA. They live in mild conditions far removed from the rigours of high mountains and the far north. They are seldom subjected to snow or frost and their roots are anchored in a good, deep soil, with the result that the trees grow immensely tall with massive trunks. Redwoods often exceed 100 metres (330 feet) in height and 10 metres (33 feet) in circumference.

French pine forest
As well as the great belts of pine forest that stretch across North America and Asia, there are many patches of pine forest in Europe, like this one in Vosges, France.

The northern coniferous forests cover a vast area, from the Pacific coasts of Canada and the USA through much of Europe and into Siberia, China and Korea. In North America and Europe they are known as the boreal forest, while in Asia they are known as the taiga, a Russian word meaning 'dark and mysterious woodland'.

The southern boundary of both the taiga and the boreal forest is marked by a gradual transition to deciduous trees. Their northern edge is bordered by vast open wastes of frozen tundra. Although conifers are well adapted to withstand low temperatures and harsh conditions, to the north is a climate which even they cannot tolerate. The gradual takeover of tundra vegetation begins with a thinning of the forest; the trees then become more spaced out and smaller, then progressively stunted and malformed. This gradual change from conifers to treeless tundra may take place over several hundred kilometres.

A similar takeover occurs with increasing altitude on mountains in many parts of the northern hemisphere and also in areas of South America, southern Africa, New Zealand and Tasmania, where the trees meet the perpetual snow line. However, in these mountainous regions the corresponding transition zone is much narrower. This 'tree line' is a meeting place, not only of the trees and tundra, but also of the snow-dwelling polar animals and those that inhabit the coniferous forests.

Despite the relative poverty of the coniferous forests, a more varied habitat for animals is provided by a generous scattering of glacier-scoured lakes and slow-moving rivers amongst the trees. The few resources available are at least provided all year round. The trees offer an ample but unvaried diet of seeds, buds, bark and young needles, which means that only specialized feeders are able to survive. Among the birds, for example, the Siberian jay (Perisoreus infaustus) is able to open the cones with its strong bill, while many members of the Galliformes (grouse, turkeys and pheasants) have stomachs which are capable of digesting pine needles.

Coping with the climate

Bad winters can pose great problems. Many animal species store food in preparation for snow; red squirrels (Sciurus vulgaris), for example, collect and hide cedar seeds and nuts in the hollows of trees, and as a special delicacy for the long, cold, dark months they impale mushrooms on the spiked ends of branches. Other species, such as brown bears (Ursus arctos) and hedgehogs (Erinaceus europaeus), let the cold months pass them by as they hide away in deep sleep, using up reserves of fat which they had accumulated in autumn when food was more plentiful. Lemmings and voles retreat beneath the snow, into a world of their own where the temperature is considerably higher than in the biting winds outside.

The long-awaited spring and summer bring relief to the beleaguered forest and, as the snows melt, young, tender shoots once again provide sustenance for herbivores such as red deer (Cervus elaphus) and moose (Alces alces).

But despite the practical problems of living in such a severe environment many animals make it their home, particularly in summer. The very isolation and harshness of the majority of coniferous forests make them among the least disturbed wildlife habitats in the world.

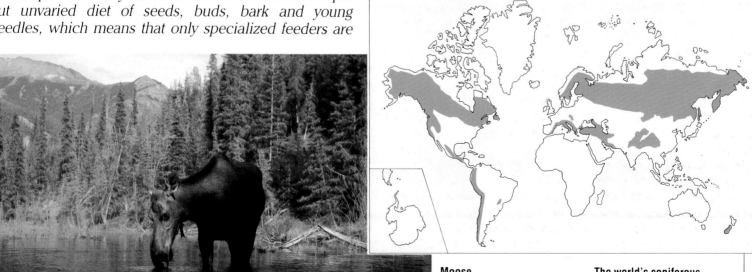

Moose
In North American woodland lakes, this impressive mammal can be found feeding on the soft water plants.

The world's coniferous woods
Coniferous woodlands stretch in a broad belt across the northern hemisphere.

Solitary hunters

Two mammals of the coniferous forest – the black bear and the wolf – are known for being loners, though in the case of the wolf this reputation is largely unfounded – the famous howl is often used as a way of keeping in touch with the rest of the pack.

American black bear

The American black bear is so variable in colour and other characteristics that scientists have distinguished no fewer than 18 different sub-species. The blue bear, the Newfoundland black bear, the Eastern black bear and many others are simply variations of the same species.

Black bears once inhabited most forested regions of North America and northern Mexico but in recent years they have declined in numbers in many areas. As with their larger relative, the grizzly (*Ursus arctos*), in most places they are treated as a game animal and sports hunting has been the main cause of death. However, compared to the grizzlies their intelligence, more secretive nature and higher breeding rate have helped them to survive over a wider range and in greater numbers. They have even begun to extend their range northwards on to the cold, treeless tundra.

Black bears are loners and usually wander by themselves in search of food. They are largely omnivorous, feeding on almost any succulent, nutritious vegetation such as bulbs, berries, nuts and young shoots. They will also take grubs, carrion, fish, young deer and even beavers, along with anything else they can find which is edible. In some places they are seen searching through rubbish tips and, particularly in National Parks, they have no qualms about raiding camp sites or begging for food from passing cars.

A single black bear will often travel over large areas – as much as 94 square kilometres (36 square miles) – if food is scarce. Males have to travel even farther in search of a mate and have been known to cover over 600 square kilometres (230 square miles) in a single season.

Grey wolf

It is possible to hear the wolf's 'lonesome howl' from as far as 16 kilometres (10 miles) away. Wolves keep in touch by howling, which may be used to call the pack together after a hunt or to warn other packs to stay away. It does seem, however, that they may howl simply for pleasure because when one wolf starts the other pack members will join in. A pack is essentially a group of five or more adult animals, usually consisting of a mating pair and all their mature offspring. The size of the group tends to increase with the size of the prey; animals as large as moose or caribou are very difficult to overpower unless there are a number of 'helpers'. The grey wolf has the greatest natural range of any living land mammal other than man and is found in coniferous forests and many other habitats all over the northern hemisphere.

American black bear *below*

ENDANGERED

Grey wolf (*Canis lupus*) *bottom right*

An adaptation for egg-laying

The great wood wasp, or horntail, is known throughout the northern hemisphere as a pest of coniferous trees. As with other wood wasps, which are the biggest and most striking of a group of insects known as sawflies, the female has a 'saw' on her egg-laying organ or ovipositor. Although quite harmless to humans and other animals, this spectacular instrument makes the wasp a fearsome-looking insect. In fact the ovipositor is used for boring through bark, so that she can lay her eggs in the new wood of trees. The larvae, which look like the caterpillars of butterflies and moths, tunnel into the middle of the tree as soon as they hatch. Here they stay for up to three years, eating the wood and weakening – and eventually killing – the tree.

The engineer of the forest

Beavers, always busy building complex dams, lodges and canals, are the 'engineers' of the animal kingdom. They are among the largest rodents in the world, well adapted to a semi-aquatic life around the streams and small lakes near undisturbed forests.

The beaver uses its paddle-like, flattened tail as a rudder, to steer right or left, or up or down, and also to give warning of danger to other beavers by slapping the water as it dives. The hind feet are webbed to propel the animal through the water, it has a sleek, waterproof coat and, as soon as it dives, transparent eyelids slide across the eyes and serve as underwater goggles. A beaver can stay submerged for up to 15 minutes, holding its breath, but most dives last only four or five minutes.

The teeth are ideal for gnawing. Beavers can fell trees up to one metre (three feet) in diameter and a 12-centimetre (five-inch) thick aspen can be cut down by a single animal in less than half an hour. As the tree begins to fall, the beaver scampers out of the way before returning to cut it into sections. The larger pieces are dragged or pushed to the water, while bundles of twigs are carried in the beaver's arms as it waddles to the water's edge on its hind legs. Most of this activity is at night, but if the beaver is particularly busy it will begin work early in the afteernoon.

Dams are built with wood, together with mud and stones; in order to make areas of still, deep water behind. This beaver-made lake must be deep enough so that the water at the bottom never freezes – even in the coldest weather – because the beavers continue their underwater life during the winter. Dams are normally a little over 20 metres (65 feet) long, though they can be up to a staggering 600 metres (2,000 ft), and are kept in repair and extended over the years by several generations.

The dams are designed to turn surrounding fields and forests into a watery world of beaver ponds in which the animals can build their houses or lodges. Surrounded by water, with their entrances under the surface, these can be as large as 12 metres (40 feet) in diameter. The floor is above water level and is bedded with dry vegetation, providing a platform for eating and drying off after a swim. All cracks are sealed with mud to prevent the winter rain from getting in.

In ideal conditions up to 14 beavers will live in a single lodge.

European beaver (*Castor fiber*) *below*
Beaver lodge in the boreal forest of Alaska *bottom*

CROSSBILL

DATA
SPECIES
Crossbill
(*Loxia curvirostra*)
CLASSIFICATION
Bird (passerine – perching bird)
DISTRIBUTION
Europe as far north as the Arctic Circle and east across Siberia and into Asia; extreme northern Africa
HABITAT
Coniferous forest, especially spruce and fir
SIZE
16.5cm (6½in)
FOOD
Seeds, mostly from the cones of pine, spruce, larch and other conifers

The crossbill is a very attractive yet slightly odd-looking bird. Its name derives from the peculiar crossed-over bill which enables it to extract the seeds – which are unobtainable to most other birds – from the cones of pine, spruce, larch and other conifers. The partially-eaten cones beneath this bird's feeding trees are often the only tell-tale signs of their presence, since the crossbill is normally difficult to spot.

There are three different crossbill species. The red (*Loxia curvirostra*) and white-winged (*Loxia leucoptera*) crossbills are found in parts of Central and North America, Europe (including Britain), the Soviet Union and Asia. The parrot crossbill (*Loxia pytyopsittacus*) lives in Scandinavia and the extreme west of the USSR.

Crossbills belong to an enormous worldwide group of birds known as the passerines, or perching birds, which includes more than half all known bird species – about 4,500 in all, encompassing swallows, wagtails, flycatchers, nuthatches and many others. They all have three front toes and one hind toe, ideal for perching.

While coniferous seeds form the bulk of the crossbill's diet, it also eats the seeds of rowan, ivy, hawthorn, weeds, grasses and thistles, as well as consuming insects such as flies and beetles. Different populations, however, have different-sized bills depending on their main food; for example, in woods of Scots pine the beaks are larger and stouter than elsewhere, in order to deal with harder cones.

Crossbills are famous not only because of their beaks and feeding habits but also because of their wanderings. Most young birds leave the breeding areas at the end of the season; normally they move only short distances before finding a suitable area to breed and feed, but when populations are high and nowhere is available they are forced to continue travelling. This sometimes results in invasions of new areas, known as 'irruptions', when phenomenal numbers of birds turn up well outside their normal range.

The crossbills' nesting season is normally very early, usually from February onwards. Females in Britain have even been found sitting on eggs at the beginning of December. The nest has a strong foundation of pine twigs with grass, wool, moss and lichen on top and a shallow inner cup of hair, rabbits' fur or feathers. A clutch of three to four bluish-white eggs with purple markings is incubated by the female for about two weeks. The young, which are born with uncrossed bills, leave after only three weeks but are dependent on their parents for another month.

CAPERCAILLIE

The capercaillie is famous for the male's flamboyant and aggressive courtship display as he tries to oust competitors and attract the favours of potential mates. He often becomes so carried away, posing and calling with his characteristic gurgling song, that he may even threaten deer, sheep or people if they happen to be nearby.

NIGHTJAR

Spring is well under way before nightjars return to Europe from their winter quarters in tropical Africa. The males arrive first, under cover of darkness, and immediately begin their characteristic owl-like 'churring' calls to establish breeding territories. When the females arrive the males increase their singing and immediately begin to display with their spectacular courtship flights.

The churring call, which often continues uninterrupted for minutes at a time, can be heard several kilometres away and is the only sure way of locating the birds. When silent and perched in the trees or on the ground they are so well camouflaged that they become virtually invisible.

Like owls, nightjars are nocturnal and locate their prey by sight. Their vision, however, is nowhere near as good as an owl's, so they hunt mainly at dusk when flying beetles, hawkmoths, craneflies and other large nocturnal insects are both active and visible. Unlike swallows and other birds that catch insects on the wing, nightjars fly with their beaks closed and only open them at the last moment to snap up the prey. It is possible to hear the snapping sound at the moment of capture from some distance.

If the weather is bad and there are no flying insects, nightjars are often forced to go without food for several days at a time. In extreme circumstances they can fall into a 'hunger coma', remaining completely motionless and using as little energy as possible until the difficulty has passed. This behaviour is unique among birds.

Nightjars nest on the ground and lay their eggs directly on suitable vegetation. If disturbed by grazing animals, people or potential predators, the female will roll the eggs to a new site a few metres away. Both parents take turns to care for the eggs and the young, which hatch after about three weeks. The young birds grow very rapidly and are able to scare away intruders by ruffling up their feathers and spitting loudly.

Once common on heathland and moorland dotted with pine trees, nightjars have declined in numbers considerably in the last 50 years. This is partly due to an increase in the number of cars on the roads. After sunset, numerous night insects are lured to tarmac roads by the warmth which they have absorbed during the day. This in turn attracts nightjars, and many of the birds are run over as they fly low or sit on the road surface. Felling of woods and increased building on heaths and commons have also contributed to the decline.

DATA

SPECIES	European nightjar (*Caprimulgus europaeus*)
CLASSIFICATION	Bird (caprimulgiformes – nightjars, oilbirds and their relatives)
DISTRIBUTION	East and southern Africa during winter; Europe as far north as southern Scandinavia in spring and summer
HABITAT	Open woodlands, forest edges and heathland
SIZE	27cm (11½in)
FOOD	Flying insects such as moths

LONG-EARED OWL

The most conspicuous feature of the long-eared owl is its 'ears', which are in fact no more than tufts of feathers – the actual ear openings are on the sides of the head. Active only at night, when it hunts for voles, mice and rats over open country or along forest edges. the long-eared owl spends the day roosting in dense tree cover. These owls, like other species in their group, have remarkably good vision even in the dark and their flight is completely noiseless, enabling them to swoop on unsuspecting prey.

NORTH AMERICAN PORCUPINE

DATA

SPECIES
North American porcupine
(*Erethizon dorsatum*)

CLASSIFICATION
Mammal (rodent)

DISTRIBUTION
Alaska to Newfoundland
and south to northern Mexico

HABITAT
Mixed forest

SIZE
Head and body length 65 to
85cm (2 to 2¾ft), tail 14 to
30cm (5½ to 12in)

FOOD
Evergreen needles and bark;
buds, roots, leaves and berries

Aristotle once wrote that porcupines can shoot their 'deadly needles like darts' over great distances at their enemies. If this was true, with more than 30,000 quills on a single animal it would no doubt be an excellent means of defence.

However, while a porcupine's quills are needle-sharp and are indeed used for defence, it cannot shoot them at all. It has been known for a loose quill to be thrown at an opponent when the porcupine shakes its body, probably because they come off very easily, though this is perhaps more by luck than judgement. Nevertheless, quills can be thrown off with such force that they become firmly embedded in trees and other solid objects.

The normal defence system is almost as impressive as the legend. Usually the quills lie flat against the body, but if an enemy such as a bobcat comes too close they are raised and spread. The animal turns its rump towards the attacker and lashes out with its barbed tail. More often than not this is very effective and the would-be predator is forced to flee.

Some species, such as great horned owls and several carnivorous mammals, have nevertheless learnt to overcome the porcupine's defence. Both bobcats and wolverines, for example, are adept at flipping them over on to their backs to expose the underparts, which are completely unprotected.

Though they sometimes forage by day, porcupines are mainly nocturnal and become active shortly after sunset. They remain in their dens (hollow logs, burrows or even crude nests in trees) during very cold or stormy weather, but they do not hibernate and are active throughout the winter, feeding on bark and evergreen needles. In summer they switch to leaves, buds, stems and fruit.

These vocal creatures have an elaborate and very noisy courtship display which uses their tremendous vocabulary of sounds to the full. Their repertoire includes wails, grunts, coughs, whines and tooth chatters. The males also perform an amusing dance which includes showering the females with urine.

A single offspring, fully equipped with hair and soft quills, is born any time from April to June. It is almost immediately able to walk about, albeit unsteadily, and can climb trees within a few days. Adult North American porcupines are able to climb up to 18 metres (60 feet) above the ground in order to obtain food. They are also able to swim well, taking advantage of their hollow quills, which are so buoyant they act as excellent lifebelts.

There are over 20 different species of porcupines falling into two main groups, one living in North and South America and the other in Europe, Africa and Asia. Only the American species climb trees; they have wider feet as a special adaptation for this, and they also have shorter spines than their Old World counterparts.

EUROPEAN WILD CAT

D A T A
SPECIES Wild cat (*Felis silvestris*)
CLASSIFICATION Mammal (carnivore)
DISTRIBUTION Parts of Europe, including Scotland and Balearic Islands, Africa, to north-central China and central India
HABITAT High woodland, particularly on the border of forested and hill country
SIZE Larger than a domestic cat, typically 60cm (2ft) long plus tail of 30cm (1ft)
FOOD Mainly mice and voles, also hares, rabbits, birds, reptiles and large insects

LYNX

The lynx (*Felis lynx*) has been exterminated in many parts of its range, because it is considered by many people to be a predator of game and domestic animals and because its pelt is of commercial value. It still lives in forested regions from western Europe to eastern Siberia and Tibet, and in parts of Alaska, Canada and northern USA. It is well suited to surviving the long, harsh winters of the northern forests, with a thick coat and furry toes to make it easier for the animal to walk in deep snow. A close relative of the lynx is the Spanish lynx (*Felis pardina*), found only in a few open forests and thickets on the Iberian Peninsula and also around the Guadalquivir Delta.

Although it is adapted to cooler climates than many other members of the cat family, the European wild cat can find it difficult to cope with the bad winter weather of its northern range. This mammal is slightly larger than a domestic cat and its heavy body and short legs make it sink deep in the snow. It also has great difficulty finding mice and other prey animals underneath.

European wild cats are found in Scotland, central and southern Europe and parts of Asia. However, their range may also include Africa and China; a very similar cat lives in these regions which some scientists believe is exactly the same species.

Until recently wild cats were thought to be pests and have been widely persecuted in many parts of their range. In fact in Britain, where they were once widespread and common, their numbers have been reduced so much that they are now restricted to the Scottish highlands. Only their secretive nature has permitted them to survive there. This is ironic because, far from being pests, they are valuable animals with an important role to play in the control of destructive rodents.

Wild cats are mainly nocturnal animals, most active at dusk and dawn, though they can sometimes be found basking in the midday sun on tree stumps or in woodland clearings. They feed mostly on birds, mice, young hares, wild rabbits and large insects such as locusts. The hunting grounds consist of meadows, farmed fields, clover fields and stubble fields around wood edges in summer, but they return to the forests in winter.

Normally these carnivores stay within their individual territories, but they do travel considerable distances during the breeding season if a partner is not in the immediate vicinity. Between two and four young are born in April or May, in a protected site carefully selected by the mother to be in a good hunting area. At first the kittens are blind and helpless but their eyes open within two weeks; they begin following their mother on hunting forays by the time they are three months old. After a training period of a further three months, their mother drives them out of her territory. She needs the land herself, to find enough food to build up the necessary fat reserves for the oncoming winter, and the young must learn to fend for themselves.

Monarch butterflies *right* (*Danaus plexippus*) gathering in Mexico
Individual Monarch butterfly *far right*

MONARCH BUTTERFLY

DATA

ENDANGERED

SPECIES
Monarch butterfly
(*Danaus plexippus*)

CLASSIFICATION
Insect (lepidopteran)

DISTRIBUTION
Western USA, Canada, Mexico,
Hawaii and Australasia

HABITAT
Overwinters in groves of trees
(fir, pine, cypress) near coast

SIZE
Head and body length 3.5cm
(1½in), wingspan up to 10cm
(4in)

FOOD
Caterpillars feed on milkweed
and other related plants; adults
take nectar from flowers

The monarch, or milkweed, butterfly is the only insect which makes long-distance, bird-like migrations twice every year. It overwinters in California and Mexico but breeds as far as 3,000 kilometres (1,900 miles) farther north.

Every year, in September, monarchs begin their spectacular southward migration. By November literally millions of these butterflies gather in the mature fir, pine and cypress forests along the coasts of California and Mexico. The same location is selected year after year, with only slight shifts in the precise trees chosen by the colonies depending on the weather conditions. We do not yet know how they locate their roosts or how the offspring return to their ancestral breeding grounds.

Once all the monarchs have arrived they crowd together in dense masses on a small group of trees, often covering every available space on the trunks and branches. The Californian colonies each contain a maximum of about 100,000 individuals, but the Mexican ones are considerably larger and may number many millions. One conservative estimate of the number of butterflies at a single site in Mexico was 14.25 million.

With such large numbers, the monarch butterfly as a species is not endangered. But all known roosts in North America are under immediate or potential threat from logging, tourism, grazing, trampling by cattle and other activities. Tourists arrive in their hundreds, even thousands, to see the spectacle but they often disturb the butterflies and make them fly unnecessarily and use up their fat reserves. The tourists also bring the constant threat of forest fires. Apart from the obvious effects of logging, if nearby trees are removed then local temperatures can drop so much at night that large numbers of butterflies die of cold. The insects counteract this problem to a certain extent by moving higher into the trees as winter progresses, but this behaviour is limited in its effect.

As spring approaches the monarchs begin to leave their roosts to take nectar, drink, soar, bask, and finally they court and mate prior to their departure in late February or March. The distinctive black- and yellow-banded caterpillar feeds on milkweed and related plants, from which it extracts a poison that remains inside its body throughout its transformation to the adult stage. This is designed to protect the insect from predators, since it is toxic to many bird species. However, orioles and grosbeaks have built up a resistance to the poison and these birds are able to feed extensively on the migrating butterflies.

PINE HAWKMOTH

Although more or less confined to coniferous forests, the pine hawkmoth is widespread and common throughout Europe, as far north as Scandinavia and northern Russia. It is one of the few moth species that have actually been able to extend their range in recent years, despite all the environmental problems caused by man in the 20th century. Most butterflies and moths have seriously declined in numbers.

In Britain, for example, in the early 1900s, this moth was restricted to the Channel Islands and a few southern and eastern counties of England. But after the First World War the newly-formed Forestry Commission began planting thousands of hectares of land with Scots pine and Norway spruce, which happen to be the main foodplants of the pine hawkmoth caterpillar. This attracted the moths to new areas and, as more trees were planted, the numbers built up accordingly. It is now found in many English counties, as far north as Birmingham. The pattern is similar in mainland Europe, where the caterpillars are so common that in some areas they have become destructive pests of conifer plantations.

Nevertheless, the adult moths do have an important role to play in the pollination of many plants. Flower species such as honeysuckle and privet rely on them almost entirely. As with all hawkmoths, they have a very long proboscis which can stretch down to the nectar deep in the base of the flowers' blossom, where no other large insect can reach. The proboscis of some tropical hawkmoth species is over 28 centimetres (11 inches) long and is used to feed from the sweet-scented, long-tubed tropical flowers which bloom at night. The moths are actually able to hover in front of the flowers, making them the insect equivalent to the hummingbirds.

Hawkmoths are renowned for the large size of both their caterpillars and the adults. There are about 1,000 species in the family, with representatives virtually worldwide, and they are all sturdy and striking creatures which seldom pass unnoticed.

Hawkmoths are among the best fliers of all butterflies and moths. Their strong, streamlined bodies are powerful enough to make their narrow wings beat so rapidly that in flight they become a blur and often make a faint buzzing sound resembling a bee or wasp.

Pine hawkmoths can be seen on the wing during June, July and August. Like most other members of the group they usually fly only at dusk or during the night and spend the daytime resting on the trunks of trees, where their superb camouflage makes them virtually impossible to see.

The moths themselves live only about a month. During the end of July and early August they lay their eggs on pine needles and die soon afterwards. The eggs hatch into attractive green-and-white caterpillars whose sole task in life is to eat pine needles. As they do so they grow rapidly and, when a satisfactory size has been reached (which is usually some time in October), they drop to the ground. There they burrow into the moss and pine needles on the forest floor and change into pupae (chrysalises). These hibernate until the following summer or, if the weather is not particularly good, until the year after that, and eventually the adult moths emerge.

DATA
SPECIES
Pine hawkmoth
(*Hyloicus pinastri*)
CLASSIFICATION
Insect (lepidopteran)
DISTRIBUTION
Throughout Europe, extending across Scandinavia into northern Russia
HABITAT
Forested areas, primarily Scots pine and Norway spruce
SIZE
Head and body length 4cm (1½in), wingspan up to 9cm (3½in)
FOOD
Caterpillar feeds on pine needles, adult on nectar of flowers such as honeysuckle

DESERTS

Unlike many other habitats such as tropical rain forests and wetlands, which are disappearing at alarming rates, deserts are not under threat. Indeed, they are actually increasing in size. Over five per cent of the world's land surface is desert already, and an area of 18.5 million square kilometres (7 million square miles) – twice the size of Canada – is likely to become so in the next two or three decades. Even the Sahara, already the greatest desert on earth, is expanding. Cave paintings by ancient tribes show that parts of it were green and fertile only 10,000 years ago, with animals such as elephants and giraffes roaming the land. During the past 50 years alone the Sahara has invaded a further 1,000,000 square kilometres (386,000 square miles) of land.

The consequences of expanding deserts, however, are as worrying as the shrinking of other habitats. The problem of desert spread is caused by the massive collection of firewood for cooking and warmth, and by over-grazing and over-cropping the land, which strips it of all vegetation. With nothing to hold the topsoil down the wind simply blows it away and the bare, hard land becomes useless. Every minute of the day, over 40 hectares (100 acres) of land that was once suitable for farming and grazing, or that once harboured a rich variety of wildlife, is destroyed in this way.

Usually we hardly notice this process of 'desertification'. Only during disastrous droughts, which give the deserts a chance to leap forward, does the problem become so serious that it is brought to the attention of the outside world. But the fact is that desertification already affects nearly 100 countries and threatens more than one-third of the earth's land surface.

Most of the world's main deserts are found in two narrow belts, one along the Tropic of Cancer and the other along the Tropic of Capricorn. Many of the desert areas are formed in the shelter of mountain ranges or in the middle of huge land masses, where there is little rain. That is not to say it never rains in the desert – but the amounts are certainly very small. In some places rain may come once or twice a year, in fierce torrents that fade almost as soon as they begin; in others it may rain only once every five or six years.

Water is also sometimes available from other sources. The Namib Desert, north of the Kalahari in south-west Africa, is unusual in that it borders a coastline. On some nights fog rolls in from the sea and as it drifts over the desert it condenses into droplets. Many animals and plants take advantage of this, among them the darkling beetles. These insects catch the drops of moisture on their legs, then lift their legs up into the air until the drops trickle down to their mouths.

It would be wrong to say that deserts are always hot. In the Gobi Desert in Mongolia, for example, night-time temperatures can fall to as low as minus 40°C (minus 40°F). Even in the Sahara, or in Death Valley in California, where daytime shade temperatures are sometimes as high as 58°C (136°F), they drop to near freezing at night. This is because when the sun goes down there is no cloud or tree cover to keep the day's heat close to the ground, so it escapes quickly and easily. As a general rule, the farther a desert is from the equator, or the higher it is above sea level, the colder it is at night.

Despite these extremes and contrasts, all desert inhabitants have two common problems to contend with: how to escape the scorching sun, and how to exist on small and irregular supplies of water.

No living thing can survive without water – indeed, one feature of a true desert landscape is the complete absence of vegetation. Yet a surprising number of species have evolved different ways of surviving on the barest minimum of moisture. In fact an astonishing variety of plants and animals live under extreme desert conditions. Many desert plants, for example, manage by living and dying in a single, often very short, growing season. Others, called succulents, store water during

times of plenty in their thick spongy leaves and stems, and use this reservoir gradually during dry periods. The cacti, which live mostly in the hot deserts of Mexico and North America, are among such drought-resistant plants. They have very long roots which are able to take up every drop of water from the ground. Some species even have poisonous substances in their leaves and roots to prevent young plants from growing nearby and competing for the water supply.

Like plants, insects also flourish when rain falls and they have also evolved various ways of surviving the dry times. Insects exist in great numbers in deserts and they play an important role in the survival of other desert animals, many of which are exclusively insect-eating. The desert locusts (Schistocerca gregaria) of North Africa and the Middle East are a good example. During dry periods they are solitary and inactive animals; but as soon as it begins to rain, and plants appear, they begin to breed. By the time the dry conditions have returned their numbers have swelled into millions and they crowd together to feed on the dwindling vegetation. The locust swarms take flight in search of food, eating any vegetable matter in their path. An average swarm is several kilometres long and can consist of 1,000 million individuals which need to eat a total of 3,000 tonnes of food each day to survive. As soon as the vegetation disappears, their numbers decline once again and they resume their solitary lives.

The soft, shifting sand poses problems for many desert animals. (Not all deserts are an endless sea of rolling sand dunes, though – some are made up of rocky hills or stony plains.) Among the sand-dwelling types are several species of gecko which have developed a kind of webbing between their toes, like a frog. Skinks do not use their legs and feet at all, they simply 'swim' through the sand by curving their body like a snake.

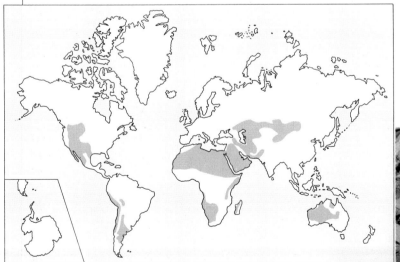

Saguaro cacti *opposite*
These imposing plants dominate the desert on the borders of Arizona and Mexico.

The world's deserts *above*
Claret cup cacti *right*
Found in the South-West USA, the flowers give these cacti their common name.

Lizards

Reptiles are probably more at home in deserts than any other group of animals. They require very little water – what they do need is often supplied in their food, so they do not need to drink – and their tough, scaly skins help them to retain moisture.

About 3,000 species of lizards are found throughout the world and many, from geckos and night lizards to chameleons and skinks, are desert-dwellers. Unlike most desert snakes, which are nocturnal, the lizards are usually diurnal (active during the day). Being cold-blooded creatures they rely on the sun to warm themselves, the surrounding air and ground in order to raise their own body temperatures.

Lizards generally try to avoid extensive fluctuations in temperature, so in the early morning they sun themselves by lying flat out in the sand while at the hottest time of the day they prefer to lie in the shade. If no shade is available they will at least raise their bellies as far as possible off the hot ground to avoid overheating. Some lizard species, such as the spiny-tailed lizards (*Uromastyx* sp.) of North America, the Middle East and parts of Asia, can survive with body temperatures as high as 47°C (116°F).

Many lizards, such as the shingleback (*Trachydosaurus rugosus*), show wonderful adaptations to desert life. This particular species not only has short legs, which it uses to 'swim' through the loose sand, but it also has transparent eyelids. This means it can see where it is going even when its eyes are 'closed' to keep out the sand.

In contrast, the strange-looking horned lizards (*Phrynosoma* sp.) of western North America are common in deserts but are not so specialized that they are necessarily restricted to this habitat. A few of the seven species of horned lizards are even found in built-up areas and gardens. In some ways this is unfortunate because their normal defence mechanism is to press flat against the ground and freeze, hoping they will not be seen; this behaviour often means that lawnmowers roll over them, with disastrous results.

The long isolation of Australia as a continent has allowed several very unusual lizard species to evolve in its deserts. The perentie (*Varanus giganteus*) is two and a half metres (eight feet) long and the world's second-largest lizard, but for all its size it is a shy desert creature. When encountered it prefers to take cover in rocks, scrubby bushes or even a hastily-made burrow in the sand. Yet when pursuing prey it is a fearsome hunter and has been reported to kill young kangaroos. The thorny devil (*Moloch horridus*) is one of the world's oddest lizards; covered in multicoloured spikes for protection, it eats solely desert-dwelling ants.

Coast horned lizard
(*Phrynosoma coronatum*) above
Lizard (*Pristurus carteri*) left

Desert scorpion

Scorpions are ideally suited to living in deserts, because they can withstand great heat. Even so, they spend most of the daytime sheltering under rocks or stones and only come out at night, dusk or dawn. As a group, scorpions are related to spiders. They are common but secretive animals, ranging in size from less than a centimetre (half an inch) to about 18 centimetres (seven inches). In total, over 600 different species of

scorpion are known.

A scorpion's prey consists of beetles, cockroaches and other small animals. When hunting it uses the sting on the end of its tail only if the prey is very large or struggles violently. The sting is also used in defence – which includes being trodden on or handled by humans. The sting ranges in seriousness from the equivalent of a bee sting to being deadly, depending on the species. Poison from the Sahara scorpion (*Androctonus australis*) is comparable in strength to that of a cobra and can kill a dog within seconds.

Scorpions have a spectacular courtship ritual. The male tightly holds his partner's pincers and walks around her in what is known as the scorpion's 'dance'. After mating, eggs are laid but the shells usually break immediately after 'birth' to release fully-developed, miniature versions of the adults. The young desert scorpions immediately climb on to their mother's back and she carries them around like this until they are either too large for her or old enough to look after themselves.

Giant hairy scorpion (*Hadrum hirsutus*) *left*

Desert fishes

The desert pupfish (*Cyprinodon macularius*) used to live in the desert zones of the southern USA, when these regions were once great lakes instead of arid sands. They are still found in some of the springs, water pools and small creeks which have been left behind, especially in parts of Arizona, south-east California and Mexico. The water in these places is typically less than a metre (three feet) deep, and is fairly cool, but the desert pupfish can live in hot springs where water temperatures reach 50°C (122°F).

The closely-related devil's hole pupfish (*Cyprinodon diabolis*) is confined to a small part of a spring-fed pool in Nevada, in the western U.S.A. The pool is about 15 metres (50 feet) below ground and the pupfish population is dependent on small animals and plants for food; these in turn are ultimately dependent on the amount of sunshine filtering down from the desert outside. In summer the pupfish population may rise to nearly a thousand but during the dull winter months their numbers drop to one or two hundred. Consequently, the devil's hole pupfish has the unfortunate title of 'most restricted fish in the world', trapped in the middle of an otherwise waterless desert with no means of escape.

ENDANGERED **Desert pupfish** (*Cyprinodon macularius*) *above*

45

SIDEWINDER

DATA

SPECIES
Sidewinder
(*Crotalus cerastes*)

CLASSIFICATION
Squamata (snakes and lizards)

DISTRIBUTION
Western North America

HABITAT
Sandy deserts

SIZE
Length 60 to 70cm (2 to 2⅓ft)

FOOD
Mostly other reptiles,
particularly lizards

As its name implies, the sidewinder literally travels sideways. With only two or three short portions of its body in contact with the ground at any one time, it literally seems to float sideways just above the ground. This sidewinding movement is an adaptation to desert life, enabling the snake to move at speeds of up to four kilometres (nearly three miles) per hour, even on loose, sandy ground. The species is capable of the usual snake-like methods of locomotion as well, but sidewinding is the only way it can reach such speeds.

Sidewinders are small rattlesnakes, but unlike many of their relatives they are not particularly venomous. Living in the deserts of Western North America, they eat mostly other reptiles and have a particular fondness for smooth-throated lizards (*Liolaemus* sp.) and whiptail lizards (*Cnemidophorus* sp.). Most desert-dwelling snakes are nocturnal and sidewinders are no exception. Unlike others, however, they do not bury themselves in the sand and are therefore seen more often.

Sidewinders give birth to fully-formed young usually between August and October. There are 8 to 15 offspring, each one surprisingly big at up to 30 per cent of the length of its parent. Birth takes up to five hours, with each young snake appearing at intervals of between 15 and 45 minutes. They quickly become expert sidewinders but are not able to breed until about three years old.

HORNED VIPER

Only a tell-tale series of S-shaped furrows in the ground give away the presence of a horned viper (*Cerastes cerastes*), one of the most poisonous of desert snakes. Adept at digging rapidly into the sand, this species spends most of its time hidden just below the surface with just a small part of its head visible. Two enlarged scales, which look like horns and give the viper its name, sit on the top of its head and protect its eyes. These scales are present in all horned vipers except, for some reason, a few populations living in southern Tunisia.

Lying in the sand cools the snake during the day and helps it to stay warm at night. There it remains – resting, or waiting in ambush for a careless rodent, lizard or bird. At the vital moment the viper unleashes itself with a spray of sand and its poisonous fangs strike deeply. If the prey manages to retreat after being bitten the snake calmly takes up its trail, sure to find the unfortunate victim in its death throes nearby.

Horned vipers live throughout North Africa (particularly in the Sahara) and Arabia. Very common, they are easily found in the vicinity of shrubs and bushes. Their scaling is much rougher than in non-desert snakes, enabling them to maintain a far better hold on the loose, sandy ground. The thick, impervious skin helps by reducing water loss. And the jagged scales along the sides of the body, when rubbed together, are used to produce a frightening sound very similar to that of a rattlesnake. This disturbing noise replaces hissing, since horned vipers (and many other desert snakes) cannot hiss: making such a sound involves the loss of water vapour from their lungs, which they can ill afford.

Desert-dwelling relatives of the horned viper include the sand viper (*Cerestes cerastes*) and the saw-scaled viper (*Echis carinatus*). Like other members of the viper group these snakes are highly evolved and extremely venomous. Their two long, hollow fangs are hinged at the front of the upper jaw and lay folded back along the top of the mouth when not required. As the snake strikes the fangs swing forwards and are driven deep into the victim, the poison being injected through the hollow core of each tooth like a hypodermic needle.

This highly efficient mode of hunting makes the vipers as a group fairly stout, sluggish snakes since they have no need of agility and speed. As they bask or lie in wait, well camouflaged against the desert rocks or scrub, they are easily trodden on or disturbed by unwary people and tend to strike back rather than flee. Coupled with their powerful venom, and the unfortunate tendency of desert walkers to discard boots and socks, this makes the vipers responsible for a high proportion of those snakebites which cause injury. The viper's venom contains a powerful nerve poison called a 'neurotoxin' which paralyzes the victim's heart and breathing muscles, in a matter of minutes. Even so, fatal bites are rare. In Australia, which is mostly desert and possesses several venomous snakes, it is said that you are more likely to be struck by lightning than be killed by a snakebite.

DATA	
SPECIES	Horned viper (*Cerastes cerastes*)
CLASSIFICATION	Squamata (snakes and lizards)
DISTRIBUTION	North Africa and Arabia
HABITAT	Sandy deserts
SIZE	60cm (2ft)
FOOD	Desert rodents, lizards and occasionally birds

ROADRUNNER

DATA
SPECIES
Roadrunner
(*Geococcyx californianus*)
CLASSIFICATION
Cuculiformes (cuckoos and turacos)
DISTRIBUTION
Southern USA and northern Mexico
HABITAT
Cactus deserts
SIZE
58cm (23in)
FOOD
Snakes, lizards, grasshoppers and a variety of other small animals

Once described as the 'early discarded doodling of a chicken designer', the roadrunner is by any standards a strange bird. A member of the cuckoo family, it is about half a metre (one and a half feet) long and spends much of its time racing around the deserts of southern USA and northern Mexico – just like its famous cartoon-character equivalent.

Like many other running birds the roadrunner has long legs with short toes. If pressed it will fly for short distances by leaping into the air and fluttering its small wings, but it prefers to keep its feet firmly on the ground. When chased by a predator this unusual bird can reach speeds of about 25 kilometres (16 miles) per hour.

The roadrunner is noted among desert birds for its expertise at killing snakes, which is done by a series of quick stabs with its long, pointed beak. It is also an avid lizard-killer. Any lizard darting from the shadow of a rock or cactus may find itself with a roadrunner in full pursuit. It gets a whack on the head as the bird overtakes and is then beaten repeatedly on the ground or against a nearby rock, before being swallowed headfirst.

Roadrunners enjoy grasshopper-hunting, often leaping about wildly to catch the jumping or flying hoppers in mid-air. They will also eat mice, birds, scorpions, centipedes and tarantula spiders if any of these are available. They are very popular on farms – not only because they help to control many pests, but also because they are such lively and inquisitive birds.

The roadrunner is not a very skilful nest-builder, making do with a loose arrangement of sticks balanced in a tree or bush. Unlike most cuckoos they incubate the eggs themselves. The baby birds are born after about 18 days, naked and with a black skin which is a heat-conserving adaptation to the cold desert nights. When they grow their feathers, these will protect them from the scorching sun and enable them to stay out in the desert unharmed throughout the day.

ELF OWL

The sparrow-sized elf owl is one of the smallest owls in the world. Only the least pygmy owl (*Glaucidium minutissimum*) of Central and South America challenges it for the title.

Elf owls are found in a range of habitats including woodland, forest, grassland and wet savannah, but they are best known for living in deserts. Confined to the south-western USA and Mexico, they are usually associated with the giant saguaro cactus which they use for both nesting and roosting. The owls depend on holes bored by other birds, such as gila woodpeckers and flickers, and sometimes even share the same cavity. The owl sleeps in the hole by day and moves out for its nightly hunting expedition when the woodpecker comes home to roost at dusk – animal cooperation par excellence.

An elf owl's clutch consists of between two and five pure-white, oval eggs, laid on alternate days during April or May. Both parents incubate for about two weeks before the white, downy young hatch, and the adult owls also feed the young birds together.

Strictly nocturnal, the elf owl rarely emerges before dusk. It uses typical owl-hunting techniques to catch crickets, cicadas, beetles, moths and a host of other animals from the ground or low-lying foliage. The only sign of an elf owl's presence is its extremely loud voice – rapidly-repeated high-pitched notes – which sounds as if it should be coming from an owl many times its size.

DATA	
SPECIES	Elf owl (*Micrathene whitneyi*)
CLASSIFICATION	Strigiformes (owls)
DISTRIBUTION	South-western USA and eastern Mexico
HABITAT	Woodlands, forests, dry grasslands, wet savannahs and, particularly, cactus deserts
SIZE	14cm (5½in)
FOOD	Large insects

SOOTY FALCON

The sooty falcon is more typically an inhabitant of small islands in the Red Sea, where it breeds in colonies and feeds on smaller birds migrating across the water. However, many individuals have adapted to living in the midst of scorching deserts in Libya and other nearby countries. The falcons treat such deserts as a kind of dry sea, catching migrant birds which are compelled to rest in places where they can easily be caught. Nearly all sooty falcons, whether living on islands or in deserts, migrate to Madagascar for the winter, where they switch to insects as their main food.

ANTELOPE JACKRABBIT

DATA

SPECIES
Antelope jackrabbit
(*Lepus alleni*)

CLASSIFICATION
Lagomorpha (rabbits, hares and pikas)

DISTRIBUTION
Southern New Mexico, southern Arizona, north-west Mexico

HABITAT
Cactus deserts and open grass-lands

SIZE
Head and body length 40 to 70cm (1⅓ to 2⅓ft); females usually larger than males

FOOD
Grasses, herbs and other plants

All rabbits and hares are experts at moving quickly and fleeing from danger. This is because their long hind legs are specifically adapted to running and leaping. They are so perfectly designed that they enable some of the larger species to reach speeds of up to 80 kilometres (50 miles) per hour.

The antelope jackrabbit, despite its name, is actually a hare; the 'antelope' is because of its tremendous leaps. Found in the deserts of southern New Mexico, southern Arizona and parts of Mexico, it is the largest member of the rabbit and hare group. There have been reports of it leaping over horses, though jumps of about a metre (three feet) high are more common.

As the breeding season approaches the male jackrabbits indulge in the mad antics common to the hare group. They chase, growl and bite, often pulling out tufts of fur or drawing blood; they rear up and box with their forelegs, and suddenly twist and deliver vicious kicks with their hindlegs. Wounds are rare, though, and usually the loser slinks away unhurt; the prize for the winner is to mate with a female.

The babies are born in an open nest called a form, lined with fur from the female's body and disguised with grass or twigs. The young can walk a few steps just after birth but they stay in the nest, freezing stock still at any sign of danger, for about a month. The jackrabbit's breeding rate is nowhere near that of their rabbit relatives although litters of eight or more young have been recorded.

As with most rabbits and hares, the antelope jackrabbit has enormous and mobile ears. In all members of the family these are important for detecting sounds which could mean danger, but in desert-living species such as the jackrabbit they have another, equally important, role. They are used as radiators, to lose heat. The ears of the antelope jackrabbit are nearly one-fourth the length of its body, and they are stuck up in the air if the animal gets really hot. In each ear there is a network of tiny blood vessels very close to the surface of the skin so that, as the air blows across, it cools the blood. If the temperature drops the jackrabbit simply reduces the flow of blood and folds its ears along its back, out of the breeze.

Another adaptation to desert life is feeding on cacti and yucca plants, which provide moisture and help the jackrabbits to avoid dehydration. They feed mostly at night, or at twilight, when it is cooler and less dangerous to move about in the open. To get past the prickly cactus spines a jackrabbit carefully chews around the base of the spine and loosens it enough to pull it away. It can then chew freely at the pulpy flesh exposed in the hole.

Although well adapted to the hot desert environment, jackrabbits prefer to avoid direct sunlight and spend most of the daytime in the shade of plants or rocks. Some species also dig burrows – an unusual habit among hares – to avoid the particularly high summer temperatures. Strictly speaking rabbits are specialized for burrowing and hares for running, but jackrabbits seem to have the best of both worlds.

GERBIL

Large numbers of rodents live in the desert. Most are nocturnal, leaving their cool underground burrows to feed only after the glaring sun has gone down. They return at intervals during the night, carrying their seeds and other food with them, to eat in the safety of their subterranean homes. The mouse-like gerbils, found in many parts of Africa and Asia, are well camouflaged and can effectively conceal themselves above ground by remaining still, their fur being the colour of the sand or ground on which they live. If they are spotted they use their powerful hind legs to escape from predators. Long, swift leaps carry them quickly away while their tails act as rudders, enabling them to alter course in mid-air.

RED KANGAROO

Famed for their jumping abilities, red kangaroos can travel at great speed in a series of leaps and bounds. A single jump can carry them an astonishing 12 metres (40 feet) through the air at speeds of 40 kilometres (25 miles) per hour.

Largest of all marsupials, or pouched mammals, red kangaroos are widely distributed throughout Australia. They are usually seen in groups of up to ten individuals, feeding on grass and other plants during the night and resting during the heat of the day. Their long tails are used for balance and as rudders when leaping, and as a third leg when sitting.

The males grow to about twice the size of females of the same age. They occasionally fight when one male's monopoly over a group of females is challenged by another. The contests may include some spectacular boxing and kicking matches but they are often resolved simply by displaying, or by locking forearms and trying to push the opponent to the ground with the hind feet, rather like a form of judo.

Red kangaroos are opportunistic breeders, waiting until the seasonal conditions are just right before mating. The young kangaroo is very small at birth, weighing just under one gram (one-thirtieth of an ounce), but has to crawl and climb all the way to its mother's pouch almost immediately. There, having attached itself to one of her teats, it remains suckling for several months before venturing into the outside world. It is another five or six months before the 'joey', as the young kangaroo is called, leaves for good.

Kangaroos live in a range of different habitats including open plains and bushy regions. They are, however, particularly well adapted to desert conditions and only need to drink during periods of extreme heat and drought. They have large ears which act as radiators for getting rid of excess body heat and they can pant and perspire to keep cool.

Kangaroos have long been hunted for their meat and hides and because they compete with domestic livestock. Hundreds of thousands are killed every year. This is of great concern to many people around the world, both because of fears that the kangaroos' future might be endangered and because of the cruelty involved.

DATA
SPECIES
Red kangaroo (*Macropus rufus*)
CLASSIFICATION
Marsupialia (pouched mammals)
DISTRIBUTION
Throughout Australia
HABITAT
Woodland edges, open grasslands and desert
SIZE
Head and body length up to 165cm (5¼ft) in the male, plus tail of 107cm (3½ft). Weight up to 90kg (200lb)
FOOD
Grasses and other plants

TUNDRA AND MOORLAND

At first sight, tundra and moorland appear to be very similar habitats. Both are usually vast tracts of open, treeless country and neither has a particularly rich variety of plant or animal life. But these obvious similarities hide some surprising contrasts.

'Tundra' is derived from the Finnish word 'tunturi', which means completely treeless. It is used to describe the enormous areas of land which in Eurasia lie north of the belts of coniferous forests, or taiga, between latitudes 60 and 70 degrees North. Tundra is also found at similar latitudes in North America, where it becomes known as 'the barren grounds', and along the coasts of the Arctic Ocean and its islands. In all these regions the average temperature of the warmest month is rarely higher than 10°C (50°F). Apart from these cold temperatures, darkness is the other difficulty facing tundra animals and plants. Over much of the region there is no sunlight for several months during winter each year, so the wildlife has to cram most of its activity into the short period of light during summer.

Moorland is found further south and does not have quite the same extremes of temperature and darkness. Unlike the tundra, which is naturally treeless, nearly level and mostly low-lying, moorland is rolling country on high ground and was once covered with pine forests and other woodland. Over the centuries – but in some cases, over little more than the past 100 years – the trees have been cleared by man and heather has replaced them. Here, the main problems are still the cold, and also the ability to retain moisture. Although there is a fair amount of rainfall, it is often quite marshy, moorland plants have difficulty in absorbing water from the acidic, poor-quality soil; and the ever-present winds on the moors add to the difficulties by increasing the evaporation of water from their leaves. The root systems of plants such as heather are consequently very deep and their leaves very tough. Moorland soil is also generally poor in mineral nutrients – especially nitrogen – and many of the plants have developed special relationships with certain species of fungi, which live among their roots and absorb atmospheric nitrogen for them.

Climate

It rarely rains as such on the tundra, but the large amount of snow the region receives is equivalent to as much as 50 centimetres (20 inches) of rainfall per year. Beneath the surface the subsoil is totally frozen, a condition known as 'permafrost', to a depth of 600 metres (2,000 feet) or more in places such as Greenland. Even in the height of the summer the topsoil thaws to a depth of only a few centimetres, or perhaps two or three metres (10 feet), and very often this is waterlogged because water cannot drain through the frozen barrier below. Nevertheless, it is this thin layer of topsoil that supports all life on the tundra.

In the far north of the tundra the icy, strong winds sweep away the scanty soil and most plants are restricted to crevices and other sheltered places. Very often, lichens are the only species visible. Farther south, particularly around the little lakes, pools and small meandering rivers, the tundra is completely covered with plants. There are no really tall trees, but knee-high forests of stunted birch and willow exist in some areas. Even the mature trees are often only a few centimetres high. The heather family is particularly common and forms an important part of the diet of many tundra animals. In the summer these plants provide a brief blaze of colour as they burst into leaf and flower, but they quickly die as the winter begins to draw nearer.

Moorland is artificially managed by man specifically to encourage the growth of heathers. Species such as common heather or ling (Calluna), bilberry (Vaccinium), and bell heather and cross-leaved heath (Erica) are the foodplants of grouse, the most important plant-eating birds of the region, which eat virtually nothing but the young shoots. Grouse shooting can be an important source of income for moorland owners and has been the main reason for preserving considerable expanses of this generally inhospitable habitat; otherwise, much moor-

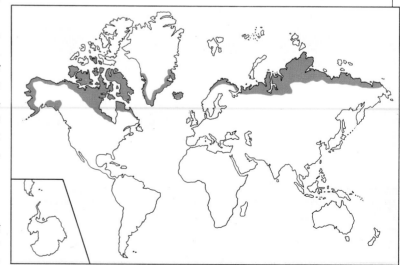

The world's tundra *above*
The tundra stretches in a belt across the northern edges of Europe, Asia and North America.

land would probably have been lost under the plough in recent years. The moors are maintained in a suitable condition for grouse by systematically burning every part once in 10 to 12 years. A region is divided into areas and each is burnt at a different stage, to leave areas of varying ages and structure. In this way the young shoots appear in recently burnt areas and provide the birds' food, while denser old clumps are used as cover and for nest sites.

There are currently one and a half million hectares (over three and a half million acres) of grouse moor in Britain and Ireland alone, mostly in Scotland. It is here, and nowhere else, that the red grouse (*Lagopus lagopus scoticus*) is found. A familiar bird in the area, it is most often seen in flight since on the ground it has excellent camouflage and so is hard to pick out in the heather. Parties of grouse often fly low and at great speed, expertly

following the ground contours. Bursts of rapid wingbeats are characteristically followed by long glides and sometimes interspersed with side-to-side rolling.

Tundra birds

Comparatively few bird species are resident on the tundra. But every spring impressive millions of ducks, geese and waders travel to the far north to breed. Most other species generally rely on the abundance of tundra insects, including mosquitoes and other biting flies, which in turn feed on the blood of birds and mammals. These insects spend the larval stages of their lives in pools and are able to survive the winter in unfrozen mud under the ice. On moorland, too, there are large numbers of small insects, notably thrips, bugs and leafhoppers, which also have piercing mouthparts but instead feed on plant sap.

The larger animals of tundra and moorland live amongst the heather, but prefer to seek out more palatable and nutritous plants to eat. They live a very precarious existence in the harsh environment and the delicate balance of predators and prey is easily upset. Populations therefore tend to fluctuate wildly between superabundance and paucity.

The harsh climate and limited resources of tundra and moorland provide a refuge for the suitably adapted remnants of communities that once lived in the last Ice Age, when cold conditions were much more widespread. The result is a rare and interesting selection of wildlife that is unique to the habitat.

Spring heath *above left*
This attractive plant, with its pale purple flowers, is common on European moorland.
Autumn tundra *left*
The Richardson Mountains in north-western Canada show how the predominant colour of the tundra can change in autumn.

Camouflage

Many animals combine their colour and shape to provide themselves with elaborate and highly effective disguises. They use almost limitless ingenuity to make themselves look like leaves, rocks or even the droppings of other animals, in order to avoid being seen by predators or potential prey. As the seasons change many species living on tundra and moorland moult their fur or feathers and grow new coats, to blend in with the snow in winter or the dark-coloured ground in summer.

Northern populations of the stoat (*Mustela erminea*) undergo a similar transformation from a summer coat of brown above and yellowish-white below to a beautiful pure white in winter. Only the black tip on the end of the tail remains the same all year round. In the southern part of its range (which includes North America from California to the Arctic, and from Europe right across to Japan) the stoat remains brown but still moults into a thicker coat for winter. Obviously warmth is more important than camouflage in some places. Some areas have both types – and the various halfway stages in between.

Varying hare (*Lepus americanus mcfarlanii) top left*
Male snow goose and **female blue goose** *top right*
Stoat or **short-tailed weasel** (*Mustela erminea) above*

The snowshoe hare (*Lepus timidus*), found throughout the tundra zones of Europe, North America and Greenland, looks just like any other hare in summer. But in mid-October, as winter draws near, it undertakes a spectacular moult and turns completely white. Also called the blue, mountain or arctic hare, this mammal has many predators to avoid, including foxes, stoats, wildcats and golden eagles. The best way not to get eaten is by not being seen, so it hides by blending in with its surroundings.

But not all pure white tundra animals are designed to match their snowy surroundings. One predominantly North American bird, the snow goose (*Anser caerulescens*), is unmistakable with its pure white feathers and black wingtips; at first glance it looks perfectly camouflaged for living in snow, as its name suggests. But there is also a so-called 'blue phase' of this species, with a white neck and head and a grey body, that becomes more common as you move from the west to the east of the bird's range. Both varieties breed in the arctic tundra; neither changes colour at different times of the year; and they often nest together in the same colonies. There is clearly more to the snow goose's white colouring than just camouflage, but exactly what, we do not know.

A moorland nest-site

One of the most beautiful and memorable sights on moorland in many parts of northern Europe is the aerial display of the golden plover (*Pluvialis apricaria*). As two or three birds excitedly chase each other at the beginning of the breeding season their characteristic 'pee-pee-yer' call can be heard from several kilometres away.

The typical moorland territory of a pair of golden plovers always contains one or more hummocks or stones near the nest, which are used as lookouts. When danger threatens, one of the adult birds tries to lead the intruder away from the nest by running forwards, to attract attention, and then running away again dragging a wing as if injured. Both eggs and young are further protected from predators by their camouflage; they are coloured to blend in with the surrounding heather and moss and are almost impossible to see. There are usually four eggs in a clutch, laid from mid-April onwards, and these are incubated by both parents for about a month. The young leave the nest another month after hatching.

In particularly favoured moorland areas there may be so many nesting golden plovers that there is not enough space for them all to breed at once. Excess pairs are forced to wait until the first in line have finished, when they can take over the vacated breeding territories. Eventually, towards the end of the summer, all adults and their young leave the moorland altogether to set up home for the winter on lowland farms, mudflats and meadows.

Golden plover (*Pluvialis apricaria*) far left

An arctic squirrel

There are about 270 species of squirrel around the world, found in virtually every habitat from dense jungles and semi-arid deserts to city parks. Although normally associated with trees, there are some species which spend their lives almost exclusively on the ground. One is the Arctic ground squirrel (*Spermophilus undulatus*), or long-tailed Siberian souslik as it is otherwise known, which inhabits the treeless tundra of the Arctic. Found in Siberia, northern Mongolia, parts of the USSR and western China, it feeds mostly on nuts, seeds, grains, roots, bulbs and other vegetation, though it will occasionally take insects and even birds' eggs.

Like most ground-living squirrels the Arctic ground squirrels hibernate during the harsh northern winters, sometimes for as long as seven months at a time. They dig burrows deep into the soil, line them with hay and then plug the entrances, to keep out unwanted intruders and the freezing wind and snow. While sleeping underground they live off accumulated fat reserves and they have plenty of food stored in their burrows in readiness for waking in the spring. However, if there is no rain or snow during their waking months and the vegetation withers, the squirrels will readily sleep again until the drought has passed.

Arctic ground squirrel (*Citellus undulatus*) below

ARCTIC FRITILLARY BUTTERFLY

The Arctic fritillary butterfly probably occurs farther north than any other species of butterfly. There is at least one record of it being found at 81° 42′N and the species rarely strays outside the Arctic circle. It inhabits mainly dry tundra and hillsides, usually at heights of 300 metres (1,000 feet) or above. Depending on the weather, individuals are on the wing for six or seven weeks from the end of June.

The Arctic fritillary belongs to a large group of beautifully-marked, strong-flying fritillary butterflies with a wide-ranging distribution. The name 'fritillary' comes from the similarity between the chequer-spot pattern on their wings and the patterning of flowers in the group *Fritillaria*, whose name is in turn derived from the Latin word for a dice or chequer-board.

The poles are not ideal places for most butterflies. Unlike warm-blooded creatures such as mammals and birds, they are unable to maintain a constant body temperature. Instead, like other insects, a butterfly's temperature fluctuates according to the temperature of its surroundings, and when it is cold the butterfly must find ways of warming itself. Many spread their wings and bask in the sun to receive the full benefit of the warming rays. Butterflies with northerly ranges tend to have darker wings and bodies than their southern cousins, to absorb more heat from the weaker sun. The Arctic fritillary is generally darker than other fritillary species of Europe and mid North America.

Fritillaries as a group tend to live in the northern hemisphere. The species that dwell south of the Equator are mountain butterflies of the African and Andean highlands. Temperatures are low here, too, and like the Arctic fritillary these insects spend a large proportion of the morning trying to warm up from the cold of the previous night. They tend to favour dark rocks for basking on, since these soak up more heat than light-coloured rocks. Once able to fly they search flowers for nectar which they suck up through their long, straw-shaped mouthparts.

ADDER

Found over much of Europe and across the USSR to the Pacific coast, the adder is a common snake in many different habitats. It occurs everywhere from bogs and hedgerows to woods and sand dunes, but prefers to live in mountainous areas and on moorlands and heaths.

Adders, also known as common vipers, are normally slow-moving animals. They frequently hunt from cover, striking at their prey (usually a mouse or a vole) as it passes and then casually following on behind until the animal dies from the venom. Only lizards are held after biting – and swallowed as soon as they stop struggling.

Adders have a very efficient poisoning system. There are two venom glands, each leading directly into the base of a long, hollow fang. When the snake strikes the fangs are embedded in the prey and the venom expelled from the fang tip, which means that it is efficiently injected deep into the animal's tissues. When not in use the fangs fold back against the roof of the adder's mouth. The venom causes bleeding, swelling and shock in the victim; most small creatures succumb within seconds and are dead in a minute.

Although they are not aggressive snakes and their bites are rarely fatal, adders are easily frightened and can be alarmed by even the smallest sudden movement. As a result, adder bites are fairly common. Man is the adder's main enemy, killing enormous numbers of them every year, though hedgehogs also despatch quite a few.

The best time to see adders is in the spring. This is when they leave their winter quarters and several often bask together in the sun. Although they love the sun and are most active during the daytime, they dislike great heat. Indeed in the hotter parts of southern Europe they live almost exclusively in the mountains, where it is cooler. In the northern parts of their range and during severe winters elsewhere they hibernate, often communally, in ready-made holes and cavities.

Adders mate soon after emergence, usually during April and early May, following spectacular 'pushing' dances by rival males competing for the females. Usually 10 to 14 live young are born, though sometimes as many as 20, each about 16 centimetres (six inches) long. As they become older the females tend to grow longer than the males and may reach lengths of 90 centimetres (three feet) or more.

Readily identified by the row of dark dia-

monds along its back, the adder – like other snakes – sheds or 'sloughs' its skin from time to time, to allow growth. Often the individual will lie soaking itself in a shallow pool for a few days beforehand (though adders, like many other snakes, are excellent swimmers). Then a fluid appears just beneath the old skin, obscuring the snake's patterning and turning it a dirty milky colour. Since the eye coverings are also affected in this way, the animal is temporarily blinded. Then its head swells and splits the old skin. As the adder slithers along, the old skin rolls back like a long sock and is eventually discarded.

Adders belong to a group of snakes known as the true vipers, which are found throughout Europe, Asia and Africa. They are closely related to the Asian and American pit vipers, such as the rattlesnakes, but lack the 'pit' (a heat-sensitive organ for locating warm-blooded prey) in front of each eye.

DATA
SPECIES
Adder
(*Vipera berus*)
CLASSIFICATION
Squamata (snakes and lizards)
DISTRIBUTION
Over much of Europe, north to the Arctic Circle, south to north-west Spain; across the USSR to the Pacific coast
HABITAT
Wide range of habitats, particularly mountainous areas, moorland and heaths
SIZE
Normally up to 65cm (26in), occasionally 90cm (36in) or longer
FOOD
Small mammals and lizards

SHORT-EARED OWL

DATA

SPECIES
Short-eared owl (*Asio flammeus*)

CLASSIFICATION
Strigiformes (owls)

DISTRIBUTION
Throughout the Old and New Worlds, between latitudes 40 and 70 degrees North, as well as in parts of South America and Pacific Islands

HABITAT
Tundra, moorland and other open country

SIZE
Head and body length 33 to 43cm (13 to 17in); weight 200 to 390gm (7 to 14oz)

FOOD
Field mice, voles and other small mammals; some birds, insects, even frogs

Together with its long-eared relative (*Asio otus*), the short-eared owl is one of the world's most effective mousers. It feeds mainly on harmful rodents; a single pair with young will eat over a thousand during the breeding season.

Short-eared owls usually hunt over moorland and other open country. They alter the size of their territories from month to month, depending on the abundance of prey, and will even become nomadic and look for new hunting grounds if food is particularly scarce. In good areas, however, there may be up to seven pairs per square kilometre (nearly three pairs per square mile). Surprisingly they co-exist with long-eared owls in many parts of their range, even though both hunt the same kind of prey. They manage to live together without too much competition because the short-eared owl hunts mostly during the daytime and the long-eared at night. They also nest and roost without disturbing each other, even though they can be as little as 30 metres (100 feet) apart, since the short-eared stays near the ground while the long-eared prefers trees.

Short-eared owls usually nest in slight hollows in the ground, preferably near a clump of vegetation and often in a marsh. The first egg is laid in late April or early May, the remainder at intervals of 24 hours. The clutch size depends on habitat, food abundance and geographical location: in tundra there are between four and seven eggs, but in Africa only two to four. In both cases, however, during years of rodent abundance the numbers increase considerably – 14 eggs have been recorded in a single clutch. In addition, although they normally raise only one brood, the owls will produce a second in response to a plentiful supply of food. The females incubate alone, starting from when the first egg is laid. Incubation takes three and a half weeks; chicks leave the nest at two weeks old.

Short-eared owls are widely distributed throughout the Old and New Worlds, between latitudes 40 and 70 degrees North. They are also found in the southern half of South America, the Galapagos Islands, Hawaii, the West Indies, the northern end of the Andes and at the mouth of the Orinoco River in Venezuela.

LONG-TAILED SKUA

Skuas breed in both the northern and southern extremes of the world, on tundra in the north and on bare ground in the south. They are renowned for robbing other birds of their eggs, young and food. A skua in hot pursuit of a tern, gull, auk or other seabird will harass it with great determination and ferocity until it regurgitates or drops its food. While the other bird gives up and flies off in fright or annoyance, the skua swoops and catches its prize while it is still in mid-air.

The long-tailed skua is the smallest-bodied and slimmest of all skuas and the one least inclined to rob other birds. Although it migrates to the open ocean in winter, it breeds exclusively on northern tundra.

The long-tailed skuas leave their breeding grounds between August and October for a completely sea-going life, feeding on small fish and surface plankton far from land. They return again to the tundra during late May and early June. Here they nest in loose colonies, laying their eggs in shallow depressions in the ground which they only occasionally line with a little dry plant material. The colonies are actively defended but long-tailed skuas rarely strike intruders, unlike some of their relatives. Arctic skuas, for instance, are particularly aggressive and will drive intruders of any kind away from their nests by dive-bombing so ferociously that their beak-stabs often draw blood.

Incubation of the eggs and care of the young skuas are carried out by both sexes. There are normally two eggs in a clutch and these hatch after about three weeks. The young leave the nest only a few days after hatching but are not able to fly for nearly a month.

Lemmings are the long-tailed skua's main food during the summer. They can often be seen hovering over their prey with a skittering, tern-like flight, the long central tail feathers from which their name is derived trailing behind. Since lemmings have cycles of abundance, skua numbers also vary considerably from year to year, according to the availability of their prey. In 'poor' lemming years they occasionally switch to feeding on berries, insects and other food.

DATA	
SPECIES	
Long-tailed skua (*Stercorarius longicaudus*)	
CLASSIFICATION	
Charadriiformes (shoebirds, gulls and auks)	
DISTRIBUTION	
Northern Scandinavia and Siberia during summer; seas as far south as southern Atlantic Ocean in winter	
HABITAT	
Open moorland and tundra during summer, offshore and coasts for rest of year	
SIZE	
Total length 55cm (22in)	
FOOD	
Mainly lemmings in northern breeding grounds; small fish and surface plankton during rest of year	

MUSK OX

DATA

SPECIES
Musk ox (*Ovibos moschatus*)

CLASSIFICATION
Artiodactyla (hoofed mammals)

DISTRIBUTION
Alaska, Canada and Greenland;
also introduced to Norway,
Svalbard and USSR

HABITAT
Exclusively on Arctic tundra

SIZE
Up to 1.5 metres (5ft) at the
shoulder; weight up to 650kg
(1,400lb)

FOOD
Grasses and sedges in summer,
mostly crowberry and willow in
winter

One of the largest animals on the tundra is the musk ox, which lives in northern Canada, Alaska and Greenland. It is a strange-looking beast with a thick coat and the longest hair of any animal in the world. Enormous furry tufts nearly a metre (yard) long on the neck, chest and hindquarters make it appear even larger than it actually is. This thick coat protects the musk ox from the intense cold which is commonplace during winter in many parts of its range. It is also useful in keeping off the mosquitoes which torment reindeer and moose in spring, and it provides some protection from predators.

Although musk ox are generally slow animals and look almost clumsy compared with many of their mountain goat and sheep relatives, they can move with surprising agility and speed when necessary. Normally, however, if threatened by predators (such as a pack of wolves) the herd bunches together in a circle or semicircle, and the adults all present their lowered heads and heavy horns to the attackers. Every so often the bulls make completely unexpected attacks by darting from the group. It is an excellent method of defence against many predators and provides a suitable fortress for the calves to hide behind, but it is quite useless against human hunters armed with guns. The species has suffered considerably from hunting in the past for meat, hides and horns, and actually became extinct in Alaska and some other areas at one time. Its range today is due only to recent reintroductions to some of the old haunts.

The name 'musk' refers to the characteristic odour that emanates from the males during the rut (breeding season). When about six to ten years old, each dominant male attempts to drive all the others away from the herd in order to take charge of the females. This involves a series of displays, threats, serious fights and sometimes head-on charges at up to 40 kilometres (25 miles) per hour. The losers usually remain nearby, on their own or in small groups, until autumn, when they rejoin the herd in readiness for winter. When all the animals are together once again the herd usually numbers around 15 to 20 individuals, though as many as 100 is not unusual.

The single young are born between April and mid-June, usually in a river valley, by a lake shore or in a meadow. In winter the herds move as far as 80 kilometres (50 miles) away, to hilltops, slopes and plateaux, where the prevailing winds keep snow depths to a minimum.

WOLVERINE

Looking like a cross between a dog, a bear and a giant long-haired weasel, the wolverine is an extremely powerful animal. Although it is occasionally killed by packs of wolves, or by a grizzly bear or puma, it makes a fearsome opponent and has been seen driving these animals from their kills single-handed. It also has tremendous stamina and can run for 15 kilometres (nine miles) or more without needing a rest.

Found in the forests, mountains and open plains of the tundra and coniferous zones in the north, the wolverine is a solitary animal except during the breeding season. There are from two to four young per litter, born in a den under the snow between January and April. For the rest of the year wolverines rely for shelter on roughly constructed beds of grass or leaves in a cave or rock crevice, or hide under a fallen tree or in a disused burrow.

The wolverine is also known as the 'glutton', because it was once believed to eat ravenously until its stomach would expand no further. It eats a range of food including carrion, the eggs of ground-nesting birds, lemmings and berries. Sometimes in the winter it also takes larger animals such as reindeer, roe deer and wild sheep. Its large snowshoe-like feet mean that it can run faster on the soft snow than many of these animals, which therefore make easy prey. Food left over after a meal is covered with earth or snow, or wedged in the fork of a tree, for another day. A wolverine may eat such stored provisions as much as six months later.

Although mostly nocturnal and rarely seen in its remote and wild home, the wolverine is an unpopular animal. It has learnt to search for hunters' traps and eat the animals it finds caught in them. It occasionally breaks into cabins or food stores and sprays their contents with its strong scent. It has even been accused of preying on domestic reindeer. As a result, wolverines have for a long time been intensively hunted. Their fur is quite valuable, and is used by people living in the Arctic for parkas since it accumulates less frost than that of other species.

Where human populations are densest, wolverine numbers have declined dramatically during the past century. A single animal's home range can be enormous – often as much as 2,000 square kilometres (770 square miles) in winter, about a quarter of which is defended territory – so nowhere is the population very dense. It is now a rare animal in the USA, has virtually disappeared from Canada and is restricted to only small parts of its original range in Scandinavia and the northern part of the USSR.

DATA
SPECIES
Wolverine
(*Gulo gulo*)
CLASSIFICATION
Carnivora (carnivore)
DISTRIBUTION
Parts of Scandinavia, northern USSR, USA and south-central Canada
HABITAT
Tundra and coniferous forest
SIZE
Head and body length up to 1 metre (40in), tail length 17 to 26cm (6½ to 10in)
FOOD
Varied, including carrion, eggs, lemmings, deer and berries

LEMMING

The Norway lemming (*Lemus lemus*) is probably the best known of the nine species of lemmings. Found in Norway, Sweden, Finland and the extreme north-west of the USSR, it is famous for its tremendous population fluctuations and mass migrations. Lemmings feed on sedges, grasses, berries, lichens and various other plant food. When there is a good supply their populations build up to enormous numbers; eventually there are so many that they eat all the food and are forced to leave to find new pastures. They do not commit suicide, but many die by attempting to cross rivers.

GRASSLANDS

More than one-fifth of the total land area of the earth is made up of grasslands of one kind or another. Occupying the vast areas between forests and deserts, grassland habitats occur in many forms, each with its own distinct character: the rich veldt in South Africa; the vast savannahs of East Africa; the rolling prairies of North America; the dry chacos and pampas of South America; the desolate steppes of Central Asia; the lush meadows and chalk downlands of Europe; and the semi-arid grasslands of Australia.

The grasses were among the last groups of flowering plants to evolve, but they have more than made up for lost time and are now the dominant species over much of the world. They can tolerate hardships that would kill or cripple other plants, from fire and drought to grazing by animals and cutting by lawnmowers.

Many of the world's 10,000 grass species are specially adapted to cope with fire damage, which is an ever-present hazard in the dry season. Their growth is often more extensive under the ground than above, and their root systems persist even when the hazard has passed. Drought-resistant shrubs and trees, which grow where there is little grazing, particularly along rivers and streams, can withstand the dry season but not the fire, and so the flames can actually help grass to compete with these larger plants.

Many of the wild animals that are well adapted to living on grasslands are actually important in maintaining them. If sheep and rabbits did not continually eat new shoots, many chalk grasslands in southern England would quickly be taken over by scrub. The vast herds of bison (Bison bison) that used to roam the prairies in North America also cropped the grass and stopped regeneration of trees by eating the seedlings.

Grasslands harbour a bewildering number of animals. Some of the world's largest species, such as the giraffe (Giraffa camelopardalis), elephant (Loxodonta africana) and black rhino (Diceros bicornis) live on the savannahs in East Africa. Each of these, and other, African grazers has its own preference for particular plants, or parts of plants, so they rarely compete for the same food. Wildebeest (Connochaetes taurinus), zebra and eland, for example, can happily live side by side in great numbers, putting the vegetation to thorough – but not excessive – use.

Since plant-eaters always have to be wary of predators, grazing in mixed herds is useful for safety reasons. Each animal is able to use its own expertise in vigilance to

spot potential danger, which complements that of others. Without such a combined effort avoiding predators on the vast open plains, where concealment is difficult, would be far more difficult.

Great numbers of African birds also accompany the grazing animals. Among those which live communally are storks, egrets and the ostrich, the largest bird in the world. Other species, such as the kori bustard and the strange-looking secretary bird, prefer to live alone but are a common sight striding through the grass.

In Australia the majority of grasslands are largely inhospitable to man. Yet they harbour many of that country's most distinctive animals – including kangaroos, emus, kookaburras and spectacular flocks of parrots, cockatoos and budgerigars.

Most of the large animals that once inhabited South America's grasslands have since disappeared. They were unable to survive the more advanced carnivores which travelled southwards over a million years ago, when North and Central America became joined to South America. But over the years the carnivores themselves could not adapt to the new conditions successfully and they, too, eventually died out. The only South American grassland mammal of any size left today is the pampas deer (Blastoceros campestris), while the largest inhabitant of all is a bird, the ostrich-like rhea (Rhea americana).

But South America still has plenty of smaller mammals, such as giant anteaters (Myrmecophaga tridactyla), armadillos and the pampas fox (Dusicyon gymnocerus), as well as the usual myriad of grassland invertebrates. Most grasslands around the world have their own miniature jungle of tangled roots, matted stems and clumps of growing leaves, each with its own community of tiny inhabitants: ants, termites, grasshoppers, earthworms, beetles and many others.

During the last 10,000 years vast areas of land have been artificially converted to grassland by man, usually at the expense of natural forest. This is because grasslands are vital to the survival of people the world over. Many of our food crops are grasses, including maize, wheat, barley, rye and millet. Most of the meat we eat and the milk we drink comes from animals fed on grass, while even sugar and rum are made from a grass – sugar cane.

But humans have also been destroying grasslands in many parts of the world. Overgrazing and overcultivation have led to erosion, and grasslands have been turned into deserts throughout Africa and in many other places. In North America bad farming practices led to the infamous dust-bowl disasters in the mid-west during the 1930s. Elsewhere, thousands of square kilometres of grassland are transformed into concrete and tarmac every year as our cities and roads proliferate.

The result of this mismanagement of grasslands is hunger and starvation, which is being felt all over the world. The solutions are to improve productivity by increasing yields, to develop new varieties of cereal which can survive droughts and combat pests, to reduce levels of grazing and cultivation so that the land can sustain them, and so on. But to do many of these it is essential to have a satisfactory reservoir of wild grasses which are resilient in the face of environmental change, or continuously evolving to cope with new conditions, for future crop breeding. And it is essential to manage existing areas carefully, rather than relying on the destruction of forests to create more grasslands, as we seem to be doing at the moment.

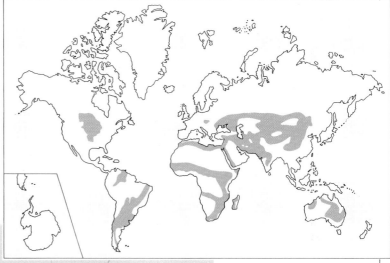

The world's grasslands *above*
Grasslands occur in great belts across the world, from the Argentine pampas and African savannahs to the steppes of Asia.
Kenyan grassland *left*
This type of scene, with the grassland punctuated by trees, is typical of this area.
'Cerrado' vegetation *opposite page*
These flowers and grasses occur in Brazil.

Termite hills and ant nests

Termites and ants are the master-builders of the insect world. They are social animals, living in colonies of up to ten million individuals packed together in their own iron-hard, sealed mounds or excavating tunnels deep under the ground. In many parts of the world grassland and savannah landscapes are literally dotted with termite hills and ant nests.

The largest termite mounds tower more than seven metres (23 feet) into the air, and the ground beneath is often honeycombed with chambers and passages. The mounds are highly organized dwellings with central heating and air conditioning systems, a royal chamber, living quarters, food-growing areas, and even their own armies of soldiers which fight off potential intruders with their big heads and powerful jaws.

Despite the termites' armoury and protection, many animals have taken advantage of the rich source of food in termite hills and also in sizeable ant nests. The spiny anteater or short-beaked echidna (*Tachyglossus aculeatus*) eats ants and termites and virtually nothing else. Instead of having teeth, it uses a pad of horny spines on its tongue to grind the unfortunate insects against similar spines on the roof of its mouth. Echidnas are found in Australia, Tasmania and Papua New Guinea and are one of only two mammal types that lay eggs (the other being the duck-billed platypus, also from Australia).

The giant anteater (*Myrmecophaga tridactyla*) of Central and South America must eat more than 20,000 ants each day to survive. It usually ignores termites, army ants and other species with large jaws, but it can devour 200 individuals of less harmful species in under a minute. Found east of the Andes, and as far south as Uruguay and north-western Argentina, the giant anteater locates new ant colonies by scent and quickly breaks them open with its powerful claws. It travels from colony to colony, taking only a small number from each to avoid over-exploitation of its food, and picks the ants up with its tongue, which is 60 centimetres (two feet) long. The ants get caught on the minute, backward-pointing spines and in the anteater's sticky saliva as the tongue flicks in and out up to 150 times a minute. Even hiding deep in their nest tunnels, the ants hardly stand a chance.

Echidna or spiny anteater (*Tachyglossus aculeatus*) below

ENDANGERED

Giant anteater (*Myrmecophaga tridactyla*) centre

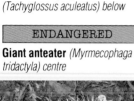

Dung beetle

The dung beetle or tumblebug (*Scarabaeus sacer*) has the amusing habit of rolling balls of dung, each many times bigger than itself, with its back legs. A natural scavenger of southern Africa's veldt, it rolls the ball of dung to a suitable soft spot in the ground and then buries it a few centimetres below the surface. There it keeps its 'larder' safely out of reach of other dung-eating beetles and consumes it over a period of days or even weeks.

The female dung beetle also lays her eggs on the nutritious balls, one egg on each, so that when the white grubs hatch out they have plenty to eat until they change into adults and can roll their own.

Spring hare

The spring hare (*Pedetes capensis*) is a common animal throughout the savannahs of southern and East Africa. Although entirely unrelated to kangaroos it looks remarkably similar as it leaps and hops around on its long hind legs, using its tail for balance.

Spring hares are small enough to be killed by snakes, owls and mongooses, yet large enough to be worthwhile prey for lions, people and other sizeable predators. They are therefore highly specialized for detecting danger and fleeing. A single leap can carry them four metres (13 feet) or more when being chased and they take many other precautions – just in case.

For instance, they are active only at night, when it is safer to forage for grasses and insects, and even then they stay fairly close to their burrows. They also feed in groups of about six because a group is more efficient at detecting predators than an individual. And, if they do have to dash down their burrows, once inside the spring hares often block the entrances with soil as a last-ditch effort to stay alive.

American bison

Few animals through history illustrate the destructive powers of man better than the American bison (*Bison bison*). Experts estimate that in 1700, before the 'white man' arrived on the prairies, more than 60 million bison wandered over North America. In some places there were single herds of more than a million animals. For generations many Indian tribes had depended largely on these animals for their livelihood, killing small numbers without affecting the overall populations.

But as soon as European settlers arrived on the continent they began to kill enormous numbers of bison – to deprive the Indians of their wild herds, to free the land for farming, and for sport. The legendary 'Buffalo Bill' Cody claimed to have killed 4,862 buffaloes in one year and literally millions of hides were sold to American dealers in the period from 1850 to 1880. A couple of years later the bison was extinct in the wild.

Fortunately, a few were kept in captivity. The species has just managed to survive, thanks to an extensive captive-breeding programme. Their numbers now stand at between 30,000 and 50,000 but nearly all of these animals live in parks and refuges where they are comparatively safe from the ravages of hunting.

BANDED MONGOOSE

SPECIES
Banded mongoose
(*Mungos mungo*)

CLASSIFICATION
Carnivora (carnivores)

DISTRIBUTION
Africa, south of the Sahara

HABITAT
Grassland, bushland, woodland
and open rocky country

SIZE
Head and body up to 45cm
(18in), tail up to 29cm (11½in)

FOOD
Beetles, millipedes and other
small animals

A small group of 20 or so banded mongooses (*Mungos mungo*) foraging on the savannah is a common sight in many parts of Africa. Poking their noses into holes, overturning rocks with their paws and scratching the ground with their sharp claws, they spend much of the daytime searching for beetles, millipedes and other small animals to eat. They have excellent eyesight, hearing and smell and always hunt together, often travelling several kilometres in a day.

When individuals disappear from sight, behind a rock or bush, they keep in touch with their fellows by making continuous birdlike twittering sounds. They also have a special alarm 'chitter' to warn one another if danger is nearby. Their main enemies are hawks, eagles and large snakes. A parent will often fight off such predators if its young are in danger.

The mongoose is famous for its prowess at killing snakes. Its reactions are so fast that it can dodge each time the snake strikes, continually provoking it until it tires so much that the mongoose can grab and kill it.

At night banded mongooses return to their dens. These are mainly in old termite mounds but hollow logs, abandoned aardvark holes, rock crevices and a variety of other places will do. A typical den consists of up to nine entrance holes, a large central sleeping chamber and several smaller chambers. It is usually kept only for a few days but favourite sites may be inhabited for up to two months.

More time is spent in the den when young are present. Breeding is synchronized within the group so that all the females bear kittens at approximately the same time. They usually breed several times a year and have two or three young per litter. The kittens are kept together and raised communally by the group, being guarded by one or two adult males while the others forage.

The young begin to travel with the adults when only one month old. The group's home range varies considerably to a maximum of about 400 hectares (1,000 acres) which was recorded in the Serengeti, in Tanzania. Ranges often overlap and intergroup encounters, with much noisy and hostile fighting and chasing, are fairly common.

RATTLESNAKE

Anyone who has heard the chilling sound of a rattlesnake can imagine how effective it is as a warning to intimidate and frighten away enemies. When disturbed or coiled ready to strike, the snake holds the tip of its tail upright and vibrates it rapidly to produce the rattling sound. If the intruder does not heed the warning, which can be heard up to 30 metres (100 feet) away, the 'rattler' can strike from over a distance of a metre with lightning speed. Ironically, although rattlers can perceive vibrations in the ground they cannot 'hear' as we do – so they cannot hear the rattling themselves!

All 30 species of rattlesnake have a rattle at the tip of their tails. It consists of hard, dry, chainlike scales which are loose but do not normally break off during moulting. Contrary to popular opinion, the number of rattles does not indicate the snake's age; after it has developed six or eight scales the older ones begin to break off, so it is impossible to tell how many there have been.

The rattlesnake's sinister looks, with its large yellow eyes, vertical slit-like pupils and protruding eyebrows, give this reptile a bad reputation.

Particularly in the north, where they gather together in large groups of up to 1,000 to hibernate, large numbers are killed by local people. Throughout their range, from Canada to Argentina, they are just as unpopular. But in truth rattlesnakes prefer to hide inside crevices and under shrubs, where they can avoid contact with people. Their food is mainly mice, voles, rats, chipmunks and other small animals.

Male rattlesnakes have ritualized fights to win females. The contenders lift their bodies nearly a metre in the air and wrap themselves around each other, moving in a swaying motion until one of them gives up and leaves. All rattlesnake species do not lay eggs but bear live young, usually between August and October, though often a few months earlier in the north. Birth takes up to five hours, with the young animals appearing at intervals of 15 to 45 minutes. There are between eight and 15 offspring in all, each up to a third the length of their parents.

DATA	
SPECIES	Western diamond-back rattle-snake (*Crotalus atrox*)
CLASSIFICATION	Squamata (lizards and snakes)
DISTRIBUTION	South-west and central USA; northern Mexico
HABITAT	Dry, rocky and shrub-covered terrain
SIZE	1.5 to 2.2. metres (5 to 7¼ft)
FOOD	Rodents and other small mammals

BOOMSLANG

Although it has a very potent venom, the boomslang, or common African tree snake (*Dispholidus typus*), is not considered to be very dangerous to man. It is a shy and retiring animal and its fangs are too far back in its mouth to be used for injuring or killing large creatures such as potential predators. As with other 'rear-fanged' snakes, the fangs are specifically for injecting poison into small animals as they are about to be swallowed. Human deaths and accidents have occurred, but mostly from people trying to capture or handle the snakes without due caution.

The boomslang is an agile tree-dweller of savannahs and other open country in the Orient, Africa and North, Central and South America.

gazelles and antelopes, was gradually pushed out and the lions were forced to hunt domestic animals to survive. The farmers immediately began to trap and poison them and, as their animals ate away the natural undergrowth which the lions need for stalking, the chances of the lions catching wild animals became progressively smaller.

The conflict was potentially disastrous. Eventually, it was solved by building a stone wall around the area and gradually moving the farmers to suitable areas nearby. This seems to have worked and the lion population, although still precariously small, is gradually increasing.

How are such small populations of animals counted? In the Asiatic lions' case, surveys were devised based on their feeding methods. One technique was to record the number of kills assumed to be made by lions; a second was to count pawprints or other tracks. A third method was to use buffalo calves as 'bait' – rather distasteful, perhaps, but deemed necessary in such an extreme case. It was thought that most of the lions would find a bait in a day or two and stay with their kills for another two or three days, so allowing their minimum numbers to be assessed.

When these census techniques were tried at different times in the Gir Forest some years ago they yielded numbers of 160, 166 and 162 individuals. The close agreement between methods was taken as a sign of accuracy and the true severity of this animal's predicament was made startlingly obvious.

Lions once roamed vast areas of Europe, Asia and most of Africa. Those in the Gir Forest are now the only ones found outside Africa, where they have also declined in numbers in recent years. The Asiatic lion is similar to its African cousin in size and general appearance but it has a slightly heavier coat and smaller mane. There is also a prominent ridge of fur and skin along its belly.

Both forms hunt in a similar way, alternately creeping and freezing, and utilizing every available patch of cover. Two or more lions often approach the prey from opposite directions, so that if one is seen as it makes its short dash then another can ambush the victim as it attempts its escape. Lionesses do most of the work and usually make the kill but then stand aside to let the males eat first – hence the saying 'the lion's share'. The females eat next and the cubs last, which serves as a natural population check when food is scarce.

DATA	
ENDANGERED	
SPECIES	Asiatic lion (*Panthera leo persica*)
CLASSIFICATION	Carnivora (carnivores)
DISTRIBUTION	Gir Forest in western India
HABITAT	Savannahs and sparse woods
SIZE	Head and body length in male up to 2.5 metres (8ft); tail length up to 1 metre (3ft); females smaller
FOOD	Wild gazelles, deer, pigs and other medium-sized animals

ASIATIC LION

Until the last century, the Asiatic lion was fairly common in savannahs and sparse woodlands over much of Asia. But many were killed for sport and others came into conflict with farmers. By 1940, these lions were extinct everywhere except one place: the Gir Forest in India. Today, the 200 Asiatic lions still alive dwell in that same locality.

The Gir Forest is a region of about 1,300 square kilometres (500 square miles) of dry, rolling hill country. Although it is a reserve, the lions had to share it until recently with several thousand farmers and more than 50,000 domestic cattle, oxen, water buffalo, camels and goats. The wild food of the lions, such as

CHEETAH

The cheetah is the fastest land animal, able to reach astonishing speeds of 110 kilometres (70 miles) per hour or more. But it tires very quickly and can run at top speed for less than 300 metres (1,000 ft), so most of its intended victims get away.

Cheetahs hunt during the day, standing on a fallen tree or termite mound to watch the surrounding grasslands for small mammals such as antelopes or hares. As soon as potential prey is spotted – and cheetahs can see extremely well – they begin to stalk. Their spotted coats blend in with the tall, dry grass of the plains and make them extremely hard to see as they creep closer and closer. Only when they are within a hundred

metres or so do they make a lightning dash. In a successful hunt, the cheetah knocks its victim down with a paw and then bites its throat. It then drags the dead animal away to eat in a shady spot, away from hyenas and vultures.

The cheetah is usually a solitary animal, though males occasionally travel in small groups of two or three, while females can often be seen with their cubs. There are three cubs in an average litter. At first they huddle close to their mother for protection and to shelter from the tropical sun. She nurses them for between three and six months, until they are able to eat meat. From then on she must kill a gazelle or a similar-sized animal every day to meet their constant demands for food. As soon as the cubs leave to fend for themselves the mother reverts to hunting every two to five days, which is normal for a cheetah living on its own.

The young cubs spend most of their early lives playing games: mostly stalking, swimming and pouncing, in preparation for adult life. Unfortunately nothing will prepare them for the ever-increasing presence of people. Adult cheetahs are hunted for their spotted coats and are far less adaptable than, for example, leopards, which often adapt to towns and villages. Cheetahs are still widely distributed in Africa but as a result of hunting are now rare in many parts of their range; in India and the Middle East they are virtually extinct.

DATA
ENDANGERED
SPECIES
Cheetah (*Acinonyx jubatus*)
CLASSIFICATION
Carnivora (carnivores)
DISTRIBUTION
Africa, a few left in the Middle East and possibly Asia
HABITAT
Varies from semi-desert and open grassland to thick bush
SIZE
Head and body length up to 1.5 metres (5ft), tail up to 80cm (2½ft)
FOOD
Gazelles, impalas and other medium-sized animals

AFRICAN ELEPHANT

DATA

ENDANGERED
SPECIES
African elephant
(*Loxodonta africana*)
CLASSIFICATION
Proboscidea (elephants)
DISTRIBUTION
Africa south of the Sahara
HABITAT
Savanna grassland and forest
SIZE
Males over 3 metres (10ft) high at the shoulder, weight up to 6,000kg (13,200lb); females considerably smaller
FOOD
Grasses and some leaves in the wet season; twigs, branches, bark, flowers, fruits and roots in the dry season

In 1970 there were about 3.3 million elephants roaming Africa. Today there are less than one million. In the last few decades enormous numbers have been killed by poachers for their tusks, which are carved and sold as ornament in the Far East and other parts of the world. Even now, with official protection and enormous parks and reserves set aside for them, between 50,000 and 150,000 African elephants are killed each year. They are also faced with continuing loss of their habitat in many parts of Africa, and their numbers are declining still more dramatically as a result.

The elephant's tusks, which occur in both males and females, are enlarged teeth. They first appear when the animal is about two years old and grow throughout life. A bull elephant that has been able to escape poachers may have tusks each weighing over 60 kilograms (130 pounds) by the time he is 60 years old, a good age for an adult. There are records of very old bulls with 130-kilogram (280-pound) tusks as much as three and a half metres (11½ feet) long, but these days such animals are very rare; in fact, half the wild elephants in Africa are killed before they are 15 years old.

The tusks are used mainly for feeding – prising bark from trees and digging for roots. They also function as weapons in social encounters. The African elephant is the largest living animal on land and its great size and strength make such encounters extremely dramatic. In addition to their tusks, elephants have large grinding teeth. Each one is used until it wears out, when it is replaced in the jaw by another.

The trunk, which is really an extended and muscularized upper lip and nose, is also used for feeding. An elephant cannot reach the ground with its mouth because its neck is too short, so the trunk is useful for reaching down as well as for reaching up into trees and shrubs, and for breaking branches and picking leaves, shoots and fruit. This versatile organ has many other uses including smell, touch, drinking (water is sucked into the trunk and then squirted into the mouth), throwing dust over its owner, and amplifying calls. It can even serve as a snorkel when the elephant is walking underwater in a river or lake.

Elephants cannot go for very long without water. They need at least 70 litres (15½ gallons) each a day, so in times of drought they dig holes in dry river beds with their trunks and tusks, until they reach the water table.

Water and food availability are important factors in determining the size of an elephant herd's home range. Bull elephants tend to live apart from the main herds, but a family unit of females (usually either sisters or mothers and daughters) and their young may cover a vast range of up to 1,600 square kilometres (620 square miles).

BANDICOOT

There are 19 species of bandicoots, found all over Australia and Papua New Guinea. Related to kangaroos and koalas, most bandicoots are rabbit-sized or smaller and spend their time poking into crevices and digging with their powerful claws, looking for small animals to eat.

Bandicoots are marsupials, which means the female raises her young in a pouch on her belly. The bandicoot's pouch opens backwards; it extends forward as the young grow bigger and contracts again after they have left. Bandicoots probably have the shortest gestation (pregnancy time) of any mammal – as little as 12½ days from conception to birth. The young are tiny when first born, roughly one centimetre (two-fifths of an inch) long and weighing about 0.2 grams (less than one-hundredth of an ounce). There are normally between two and four offspring and they all have to crawl into their mother's pouch where each attaches to a teat. They leave the pouch after about seven weeks and are weaned 10 days later.

As one litter leaves the pouch the next is being born. Bandicoots are unique in having so many young and are able to breed at this rate for as much as eight months of the year.

Despite such an incredible reproductive rate, Australian bandicoots in general have suffered one of the greatest population crashes of any group of mammals on the continent. All the species of semi-arid and arid areas have either become extinct or have been reduced to a few remnant populations. Competition with cattle, sheep and rabbits, as well as introduced predators such as foxes and cats, are to blame.

The pig-footed bandicoot (*Chaeropus ecaudatus*) is one of the most unusual of the group, with particularly long limbs and ears. The last reliable sighting of this species was in 1926, in an area south of Lake Eyre in southern Australia; it may well be extinct. It used to live on dry grasslands along with several other species such as the western barred bandicoot (*Perameles bougainville*) and the desert bandicoot (*Perameles eremiana*), which is also almost certainly extinct.

DATA
ENDANGERED
SPECIES
Pig-footed bandicoot (*Chaeropus ecaudatus*)
CLASSIFICATION
Marsupialia
DISTRIBUTION
Central and south Australia; may be extinct
HABITAT
Grassland, woodland, heath, semi-arid and arid plains
SIZE
Head and body length up to 25cm (10in), tail up to 14cm (5½in)
FOOD
Mostly soil invertebrates, also fruits, seeds and plants

THE POLES

The wild beauty of the polar regions is one of the most awe-inspiring and unforgettable sights on earth. There is something unreal about the endless views of pristine white snow, blue cascading ice and the sea, broken only by mountains that seem deceivingly close. The air is so clear that a mountain which appears to be no more than 50 kilometres (30 miles) away may be five times as far.

At neither the North nor the South Pole does the temperature rise much above 10°C and, particularly in the winter, it often falls far below this. Not only is it extremely cold but there is also prolonged darkness for almost half the year, which is the price paid for the midnight sun during summer. Since the earth's axis passes through both poles, when one is tilted slightly towards the Sun it has weeks of perpetual daylight while the other is tilted away into an endless winter's night.

Both polar regions are always cold for the same reason – they receive their sunlight more obliquely than any other part of the world, and the heat received from the Sun is proportional to the angle at which its rays strike the surface. The cold is intensified because ice acts as a mirror that reflects the solar heat energy back into the upper atmosphere.

Apart from these common climatic characteristics, the northern Arctic and the southern Antarctic are really very different. The Arctic is almost entirely frozen ocean surrounded for the most part by land; the Antarctic, on the other hand, is a land mass which is covered by a great raft of ice, in places up to four and a half kilometres (nearly three miles) thick.

Antarctica was once a much warmer place. It lay much closer to the equator and was populated by dinosaurs that roamed its forests. But between 100 and 200 million years ago the supercontinent of which Antarctica was part split up and Antarctica drifted south. Now it straddles the South Pole, stuck in the middle of a vast sea.

Although covering an area as big as the whole of Western Europe, the Antarctic has few terrestrial animals. Apart from a handful of insects (the largest of which is a wingless mosquito) most Antarctic animals live around its coast. But their sheer numbers make up for their comparatively restricted distribution. For example, it has been estimated that over 100 million birds, including penguins, albatrosses, petrels and skuas, breed around the shores and islands of Antarctica every year.

Most of these creatures get their sustenance, directly or indirectly, from the sea. The icy waters are home to enormous populations of krill, which are small shrimp-like creatures that form the basic food for Antarctic creatures such as whales, enormous numbers of fish, and the birds and seals.

In contrast, the existence of a ring of continents around the Arctic has had a great effect on its animal populations. The land has served as bridges over which many kinds of terrestrial animals have travelled northward from the warmer parts of the world. Indeed, the word 'Arctic' actually comes from arktos, the Greek word for 'bear', which is a fitting description since polar bears are one of the typical animals of the region.

Until recently, both polar regions and their wildlife had escaped the fate of other continents as man's destructive influence spread around the globe. But modern technology has allowed people to invade even these far-flung corners of the earth, and already the damage is enormous. Many populations of whales, seals and penguins have almost been wiped out; over-exploitation of krill and overfishing are jeopardizing the stability of the

entire polar ecosystems; and the pressure to exploit resources of oil and other minerals has constantly increased until it is now proving irresistible.

Underwater photography around parts of the coast of Antarctica already reveals old vehicles, fuel drums, beer cans and other rubbish. Ships routinely pump their bilges offshore and oil slicks are becoming common. But it is not only the magnificent beauty of the poles being ruined by rubbish and pollution which is of concern. There is a fast-growing realization that the conservation of their rich wildlife and natural resources are of vital long-term importance to us all.

Some scientists and fishing experts believe that krill could be one of the world's largest untapped food resources, especially as these creatures have a very high protein content. Yet already hundreds of thousands of tonnes of krill are being harvested every year, with no regard to the long-term survival of their populations. The land and air are also under threat. Dust from mining activities could lower the reflection of the Sun's rays and thereby warm up the ice cap of Antarctica. If the temperature was allowed to rise enough it could melt the ice – which contains 90 per cent of the world's fresh water – and thereby threaten coastal cities all over the world with floods.

Every year we learn more about the importance of the poles and yet every year the dangerous repercussions of commercial exploitation draw nearer. Until we can be absolutely sure of preventing such irreparable harm, we must resist the temptations.

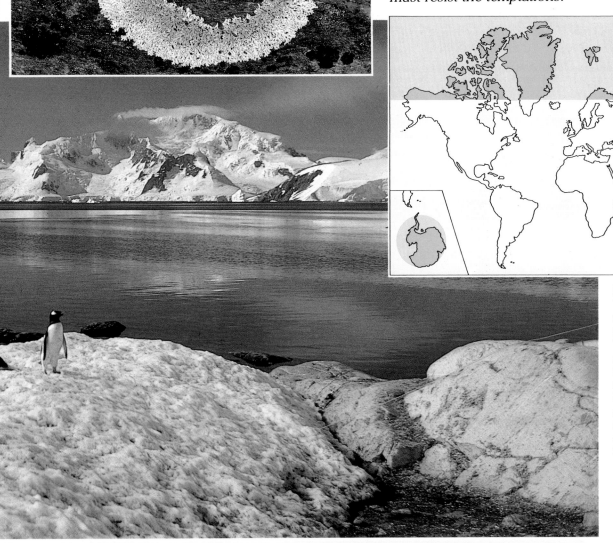

Lichen *Top*
These organisms grow very slowly and are very hardy. This particular example comes from Alaska.
The polar regions *above*
Areas north of the Arctic circle include the Northern parts of Alaska, Canada, Greenland, and the USSR. The inset on the map shows Antarctica.
Antarctic landscape *left*
A gentoo penguin stands among the ice and snow.

Seals

The seal family is a cosmopolitan group of over 30 species that include sealions, fur seals, true seals and the walrus. They are distributed all over the world, but their adaptations for life in water make them ideally suited to living in the polar regions. They have streamlined, torpedo-shaped bodies and limbs modified into flippers, and the thick layer of oily fat, or blubber, under their skin provides excellent insulation against the cold.

The crabeater seal (*Lobodon carcinophagus*) is the most abundant seal in the world and lives on drifting pack ice in the extreme south. It feeds almost exclusively on the small, shrimp-like krill which it catches simply by swimming into a shoal with its mouth open and sieving them from the water. Most feeding takes places at dusk, dawn and during the night, when the krill rise nearer the surface. Although they can be over two and a half metres (eight and a half feet) long many crabeaters fall prey to leopard seals (*Hydrurga leptonyx*), particularly as pups during their first year; many adult crabeaters are badly scarred after attacks from leopard seals.

Southern elephant seal *right*
Crabeater seal *(Lobodon carcinophagus) below*
Weddell seal *(Leptonychotis weddelli) bottom right*

tonnes. They have very fierce and noisy battles in late summer and early autumn to establish dominance over a harem of females. As they tussle they use their inflatable noses as resonators, to amplify the threatening sounds coming from their throats.

The most southerly of all seals is the Weddell seal (*Leptonychotes weddelli*), which lives in areas where the winter air temperature may be as low as minus 20°C and the sea may be frozen for seven or eight months of the year to a depth of several metres. Breeding on the Antarctic mainland and nearby islands, this seal's life is dominated by its glacial environment. The large eyes are adapted to seeing in the twilight conditions in the water beneath the ice and its front teeth protrude so that it can gnaw through to the surface to breathe. Weddell seals can stay submerged for over an hour and they can even navigate using echolocation (sounds reflected from the hard, icy surfaces) if it is too dark to see. They are the deepest divers of all seals, able to descend to 300 metres (1,000 feet) or more.

The pups of the elephant seals (*Mirounga leonina* and *Mirounga angustirostris*), found in both the northern and southern extremes, have a different hazard to face. As many as one in five are crushed to death by their huge mothers rolling on top of them, or by getting caught up in the fights between the gigantic males, called bulls. Bull elephant seals are twice the size of the females, reaching a length of six metres (20 feet) and weighing as much as three and a half

Antarctic whales

Antarctic waters offer rich feeding grounds for marine animals such as whales. A number of the world's largest whale species live there, among them blue, fin, sei, minke, southern right and humpback.

Large-scale commercial whaling first began in the northern hemisphere, in the 15th century. Despite the rich pickings, the first whalers to enter Antarctic waters did not arrive until the beginning of this century. However Antarctic whaling increased rapidly, particularly after 1926 when the first 'factory ship' appeared on the scene. Accompanied by a number of smaller catcher boats, these giant vessels could process and store the whale products – often dealing with several animals at a time – and their introduction meant that whaling could prosper quite independent on shore bases.

Hundreds of thousands of whales have been killed in the Antarctic since then. Many populations have been reduced to very small numbers, in some cases to just one per cent of their original size.

The situation has become so serious that from 1986 a total ban on all commercial whaling is planned, as a direct result of international efforts to conserve whales. But several countries are likely to ignore the ban and continue whaling, despite widespread agreement among scientists that several species are in imminent danger of becoming extinct. The future of whales in Antarctica, and elsewhere in the world, is still very uncertain.

Southern right whale
(Eubalaena australis) above
Showing mating chase.

Land animals in the Antarctic

A great variety and number of terrestrial animals live in the high Arctic all year round. But very few species are able to survive the cold, dry, windy climate of the Antarctic, with its tremendously harsh winters and lack of vegetation. Even birds and seals must leave during the inhospitable half-year-long Antarctic winter. Only a few tiny invertebrates remain behind, hidden among the sparse tangles of moss or patches of unwelcoming soil, which is really just a mixture of gravel and windblown sand. Among them are nematode worms, mites, midges and springtails – many so small that they are invisible to the naked eye. The largest permanent inhabitant of the Antarctic is a midge, only 12 millimetres (half an inch) long.

These microscopic creatures have a fascinating range of techniques to prevent their body cells freezing. Many have a special kind of anti-freeze in their blood; some lay eggs which can survive the worst periods in case the adults cannot; while others lose all their water and shrivel up when the temperatures fall, so that ice crystals do not form in their body fluids and damage them.

Springtail *(Collembola) below*

POLAR BEAR

No predator on earth approaches the polar bear in size. A typical adult male bear weighs half a tonne and may be over one and a half metres (five feet) high at the shoulder, making it twice as big as a lion or a tiger.

Polar bears live in coastal regions and on sea ice in the Arctic. They are regularly seen almost at the North Pole and some years extend their range as far south as the Gulf of St Lawrence in south-east Canada. Sometimes they wander far inland, or float many hundreds of kilometres out to sea on ice floes. They love to travel and often embark on these prodigious journeys with seemingly no particular place to go. Rarely hurried, they have a snooze if there is nothing else to do, or go for a swim, or repeatedly slide down a snow bank on their bellies.

During their travels the bears inevitably move freely backwards and forwards between Alaska or Scandinavia and the USSR, with no respect for political boundaries. In fact they are protected from hunting by a truly international agreement which has been signed by all five circumpolar nations: Canada, USA, USSR, Denmark (for Greenland) and Norway.

Today there is a total world population of 20,000 polar bears. Although this is considerably fewer than earlier in the century, when sport and commercial hunting was rife, the numbers are at last thought to be increasing, or at least stable. This is despite an annual kill of nearly 1,000 taken by Eskimos, Inuits and other subsistence hunters.

Polar bears feed mainly on ringed seals but will also take bearded seals, harp seals and hooded seals. If their normal food is unavailable they will scavenge on whale, walrus and seal carcasses and occasionally eat smaller animals. In some parts of the Arctic, notably in Churchill, Canada, they have taken to raiding dustbins and rubbish tips in towns and villages.

Cubs are born in snow dens during December and January. These dens consists of a tunnel several metres long leading to an oval chamber of about three cubic metres (yards), which is where the young bears live for the first few months of their lives. There are usually one or two cubs and they stay with their mother for up to two and a half years. Dens are sometimes excavated during the winter months, not for breeding but for temporary shelter during severe weather. These refuges are used mainly by pregnant females, which are usually the only individuals to hibernate for any length of time.

DATA

ENDANGERED

SPECIES
Polar bear (*Ursus maritimus*)

CLASSIFICATION
Carnivora (carnivore)

DISTRIBUTION
Arctic regions around the North Pole

HABITAT
Coastal regions and sea ice

SIZE
Shoulder height up to 1.6 metres (5ft); weight up to 800kg (1,750lb)

FOOD
Ringed and other seals, whale carcasses

ARCTIC FOX

One of the toughest and most adaptable of all polar animals, the arctic fox will eat anything that is available, alive or dead. It can survive temperatures as low as −80°C, swims well and can move easily over ice and snow. Individuals have been sighted over 1,000 kilometres (600 miles) from the nearest land, precariously perched on ice floes, during their seasonal migrations. They walk enormous distances as well, in any direction according to food availability and the movement of the ice.

The dense, woolly coat of the arctic fox gives it a misleadingly heavy appearance. Within the species there are two colour types: the 'white', which is generally white in winter and brown in summer; and the 'blue', which is a pale bluish-grey in winter and dark bluish-grey in summer. Although fairly common in Greenland, the blue variety makes up only a small proportion of the population elsewhere. Because the fur is so dense and attractive, arctic foxes are farmed in many parts of the world and subjected to intensive trapping in the wild.

Arctic foxes live mostly in coastal areas, though they usually have a fixed home, or den, only when rearing their young. Some dens have been used for centuries and since they are continually growing they often have over a dozen entrances and an enormous network of tunnels. As many as 25 cubs (the actual number depends on factors such as food availability and weather conditions) are born between April and July. They emerge from the den after a month or so but the parents continue to bring them food until they disperse in the autumn. They are very playful animals, racing and fighting, barking and yapping, crouching and jumping about. Their parents may bring home live rodents for them to chase and pounce on, which is good hunting practice.

In the wild most young do not survive for more than six months and few live for more than several years. Apart from trapping and the rigours of the Arctic weather they always have to keep a sharp lookout for wolves and polar bears. Nevertheless they often take risks and will follow both these animals on to the pack ice, to feed on scraps left behind after a kill.

DATA

SPECIES
Arctic fox (*Alopex lagopus*)

CLASSIFICATION
Carnivora (carnivore)

DISTRIBUTION
Northern Europe (including Greenland and Iceland), North America

HABITAT
Tundra and adjacent lands; ice-covered waters

SIZE
Head and body length 45 to 65cm (18 to 26in); tail length 25 to 45cm (10 to 18in)

FOOD
Any available meat, dead or alive

EMPEROR PENGUIN

DATA

SPECIES
Emperor penguin
(*Aptenodytes forsteri*)

CLASSIFICATION
SpheVnisciformes (penguins)

DISTRIBUTION
Coast of Antarctic continent,
neighbouring islets and southern Antarctic peninsulas of surrounding continents

HABITAT
Sea ice and snow-covered land,
adjacent waters

SIZE
Up to 115cm (45in) tall

FOOD
Fish and squid

Although there are no penguins in the Arctic, and three species which live only in the tropics, penguins are largely polar birds. The colourful emperor penguin is the largest of the 16 species in this family. It is one of the few penguins which nests entirely among ice and snow and is able to endure the coldest conditions faced by any bird. The average temperature at emperor breeding sites is minus 20°C and windspeeds often reach 75 kilometres (45 miles) per hour. During incubation several thousand birds have to huddle tightly together to keep warm. They operate a kind of rota system in order to spread the exposure to cold among the members of the 'rookery', as the breeding site is called. After a spell on the outside of the huddle an individual will force its way into the middle to regain warmth before again taking its turn at the exposed outer edge of the group.

Emperor penguins breed on sea ice or islets off the Antarctic coast. A single egg is laid at the beginning of the coldest part of the year and incubated only by the male. There is no nest or territory – the egg is simply carried on his feet, safely tucked away under a loose fold of belly skin. Meanwhile the females, which are almost identical to the males, spend the winter at sea and return just as the eggs are about to hatch. By the time they come back the males have lost almost half their body weight; before breeding they gain weight rapidly and may reach 40 kilograms (90 pounds) in preparation for their four-month incubating fast. After the breeding period they have to fatten again quickly ready for another fast, during moulting. They are often so successful at this that they can hardly waddle ashore.

The adult birds capture their diet of fish and squid with their razor-edged bills. They can dive as deep as 55 metres (180 feet) in their search for food and take squid up to one metre (three feet) long. Rear-facing, knobbly outgrowths inside the mouth and on the tongue clamp firmly on to the prey to stop it writhing and slipping away.

The chicks are fed large meals at infrequent intervals, seldom more than every three or four days. Their capacity is so astounding that they appear to be little more than pear-shaped sacks of food. They are reared through late winter and early spring and are ready for the sea by January, which is the only time of year when the sea ice is open and food is plentiful.

Like their parents, the emperor chicks are specially adapted for diving and prolonged swimming in cold water. Their flippers are flattened versions of a normal bird's flying wings, enabling them to literally 'fly' underwater, and their tail has evolved into a conical rudder. The short plumage is waterproof and windproof and they have a special layer of fat beneath the skin to keep them warm. It is not surprising that the penguin's two-legged walk and knowledgeable air makes it look very satisfied – it is very well adapted for Antarctic life.

ARCTIC TERN

With their pigeon-like flight, a gait like that of a rail, and general behaviour like a crow or gull, sheathbills are peculiar birds. They are thought to form an evolutionary link between the wading shorebirds and the adaptable gulls.

There are two sheathbill species (*Chionis* sp.), very similar in appearance and in the way they behave. Both live in the Antarctic but their ranges do not overlap. Normally they hunt among seaweed and on rocky shorelines for small animals, or steal krill from penguins before they are able to feed it to their chicks, but they have taken to feeding on the rubbish tips by Antarctic survey bases. Sheathbills have no real predators themselves – they just have to watch that their eggs are not stolen by one of their own kind.

The arctic tern is the greatest traveller in the animal kingdom. It breeds around the shores of the Arctic Ocean and in the northern Atlantic and Pacific, and then migrates halfway round the world to spend the rest of the year in the Antarctic. In this way, by staying in the north during the Arctic summer and flying south for the Antarctic summer, it enjoys more sunlight than any other animal.

On average, a single arctic tern covers up to 40,000 kilometres (25,000 miles) a year on migration alone, flying non-stop for eight months out of 12. In a typical lifetime of 25 years this is equivalent to a return trip to the Moon. Most travel by sea, feeding as they go, but overland routes are not uncommon.

Arctic tern colonies are sited on rocky islets, beaches or short turf, always close to the sea. They breed around the northern coasts of Alaska, Canada, Greenland, Europe (especially Iceland and Scandinavia) and Siberia. In Greenland they have been recorded nesting within a few hundred kilometres of the North Pole.

The birds generally pair for life. The pairing breaks down outside the breeding season and the two birds go their separate ways, but there is a strong tendency to return to previously successful breeding sites which enables former mates to rendezvous.

One or two chicks are raised from a clutch of two or three eggs. They are fiercely defended by the parents, which makes entering an arctic tern colony a very unnerving experience. The birds dive-bomb intruders, regularly striking hard and sometimes even drawing blood. Other birds, such as eiders (*Somateria mollissima*), often take advantage of this and nest in the middle of arctic tern colonies where they are protected by the terns' aggressive reaction to predators.

A very similar species to the arctic tern is the common tern (*Sterna hirundo*). It is not found near either pole but its range does overlap with that of the arctic tern in many parts of the world. In a mixed flock of the two species, seen at a distance, they are scarcely distinguishable. During the summer, the main difference is in the bills: both are bright red, but the common tern's has a black tip. Birds which cannot be distinguished by frustrated ornithologists are usually dubbed 'comic' terns.

DATA	
SPECIES	
Arctic tern (*Sterna paradisea*)	
CLASSIFICATION	
Charadriiformes (gulls, terns and skuas)	
DISTRIBUTION	
Breeds around Arctic coasts of Alaska, Canada and many parts of Europe (including Britain, Greenland and Siberia); winters in extreme southern latitudes	
HABITAT	
Oceans and arctic coasts	
SIZE	
Length 33 to 38cm (13 to 15in); wingspan up to 80cm (32in)	
FOOD	
Small fish	

KILLER WHALE

DATA

SPECIES
Killer whale (*Orcinus orca*)

CLASSIFICATION
Cetacea (whales, dolphins and porpoises)

DISTRIBUTION
All oceans of the world

HABITAT
Mostly in coastal waters and cooler regions

SIZE
Maximum weight 9 tonnes in males, 5.5 tonnes in females; up to 10 metres (33ft) in length

FOOD
Seals, dolphins and porpoises, fish, seabirds and other marine animals

Killer whales (*Orcinus orca*) occur in oceans and seas all over the world, but they prefer cooler waters and are most numerous at the poles. They are no more 'killers' than a venus fly-trap or a fox, since attacks on human swimmers and boaters are extremely rare and have always involved some kind of provocation. Far more whales are killed by people than the other way around.

Killer whales are aptly named in that they are extremely efficient predators. They travel together in groups, or 'pods', of between two and 40 animals (though up to 250 have been reported). The members stay in contact within an area of about four square kilometres (one and a half square miles). Occasionally many will join together in an attack on a larger whale or a polar bear. More often two or three individuals will band together to catch seals: one tilts a piece of floating ice by lifting it on its back, to slide the sleeping seals directly into the waiting jaws of its companions. Despite such a predatory instinct and the fact that their appetites can be enormous (the stomach of one dead male was found to contain 13 porpoises and 14 seals) killer whales unaccountably exclude man from their diet.

Killer whales, or orcas as they are often called, are easily seen and recognized by the characteristic back fin, which may be almost two metres (six feet) high in adults. Generally unconcerned by shipping and boats, they are also easily approached. Indeed they often 'pitchpole', which involves standing vertically with their head and body (as far as the front flippers) held above the surface so that they can get a good view of you.

These marine mammals breed at any time of the year. After a gestation period of 13 to 16 months a single calf, weighing about 180 kilograms (400 pounds) and two and a half metres (eight feet) long, is born in shallow water.

Although still common, killer whales are threatened because of unnecessary killing by people. In 1955 the US Navy reportedly killed hundreds of them with machine guns and depth charges, at the request of Icelandic fishermen. Today large numbers are killed because they are considered competitors of fishermen. But if it is true that there is now direct competition for dwindling populations of fish, then the source of the problem is almost certainly human overfishing.

WALRUS

The walrus is well-known for its spectacular tusks and conspicuous moustache. The tusks are extended canine teeth which grow throughout life, in both males and females, and may attain a metre (three feet) or more in length. They are used for defence, breaking through ice, as picks as the animal hauls itself out of the water, and for moving on land. They are even used as hooks, to hang on to the ice and keep the walrus' head above the surface while it sleeps in the water.

Contrary to popular belief, the tusks are not used for digging food from the ocean floor. Walruses may eat an astonishing 45 kilograms (100 pounds) of food a day, consisting mostly of starfish, sea urchins, crabs and clams. That is equivalent to about 800 large clams a day. They usually forage at depths of up to 90 metres (300 feet) in open waters near the edge of the Arctic polar ice. The bristly moustache is used as a feeling organ and acts as a filter for food on muddy ocean bottoms.

Active at night and during the day, most walruses migrate south in winter with the advance of the pack ice and then move north in spring as the ice retreats. They often ride ice floes on these migrations, simply abandoning them when the floes stray too far from their intended course.

Walruses live in mixed herds of cows, calves and bulls during the breeding season, often with over 100 individuals and sometimes over 1,000. The bulls fight one another for females, each looking for several females to form a harem. Out of the breeding season the males only gather to form large aggregations but continue to fight one another. Serious injuries, however, are rare.

The young are born between April and June. The single calf, over one metre (three feet) long, is born on the ice or on land but is immediately capable of swimming. At first, though, it travels holding on to its mother's neck – even when she swims – and is nursed for up to two years.

Practically every part of a walrus is used by Eskimos for food, material for boats or shelters, and oil or charms. The Eskimos actually 'fish' for these mammals using chunks of blubber for bait, as well as hunting them in the usual way. The walrus' other principal enemy is the polar bear, which sometimes charges at resting herds in order to try and catch calves or slower adults. An occasional predator is the killer whale; when this creature is sighted the walruses tend to

panic and in the rush to leave the water many can be injured or killed.

The subsistence hunting of the Eskimos and Chukchee Indians in historic times had little effect on walrus populations. But in the 15th century commercially-minded Europeans joined the killing and massacred large numbers, lying in ambush as the massed herds hauled themselves out of the water. Hundreds of walruses could be exterminated in an hour or two. In the 1850s as whales were slaughtered and became rare, the pressure on walrus numbers increased still further. Around 1930 there were less than 100,000 individuals, a fraction of former populations. At last conservation measures were introduced and the Pacific subspecies began to recover, but the smaller Atlantic subspecies still lags behind.

DATA
SPECIES
Walrus (*Odobenus rosmarus*)
CLASSIFICATION
Pinnipedia (seals, sealions, walrus)
DISTRIBUTION
Arctic Ocean and adjoining seas
HABITAT
Open waters near the edge of polar ice; shelters on isolated rocky coasts and islands
SIZE
Males up to 4 metres (13ft) in length; females two-thirds this size
FOOD
Clams, starfish, sea urchins and other marine invertebrates

MOUNTAINS

In many parts of the world, mountains are shrouded in superstition. Monks in Japan, for example, regard the mighty, snow-capped volcano Mount Fujiyama as sacred, and tribesmen living around Mount Kenya, in East Africa, believe their gods live near the summit, forever hidden in cloud. These attitudes are not difficult to understand. Mountainous country offers some of the most dramatic and awe-inspiring scenery to be found anywhere in the world.

Perhaps the most spectacular of all is Mount Everest, in Nepal. This is the world's highest peak, at 8,848 metres (29,028 feet). It is part of the world's greatest continental mountain chain, the Himalaya-Karakoram range, which contains an amazing 96 out of the world total of 109 peaks of over 7,300 metres (24,000 feet). Strictly speaking Mauna Kea, on the island of Hawaii, is even higher than Everest since it adds up to a total of 10,203 metres (33,474 feet) from its base on the ocean floor to its peak. A mere 4,205 metres (13,796 feet) of Mauna Kea are above sea level.

Most mountains are built so slowly that it is impossible to see any change in a person's lifetime. Volcanoes are the only exception, built with dramatic explosions that force molten rock through cracks in the Earth's crust. Some of the world's greatest mountain peaks are volcanic in origin, including Mount Kilimanjaro and its neighbour Mount Kenya.

Other mountains take millions of years to grow. Inside the earth is a hot mass of molten minerals and rocks which is constantly expanding and shrinking, due to the pulls of the Sun and the Moon and the local heating and cooling currents of molten rock. This interior is covered by a cool skin of solid rock, known as the Earth's crust. The gigantic forces of the constant disturbances within the planet cause the crust to move: some parts fold, others break up or snap, yet others wrinkle or simply rise or fall. The tops of these folds and wrinkles are mountains.

About one-quarter of the globe's land surface is mountainous. The main ranges are: the Alps and

Chamois (Rupicapra rupicapra) right These specimens are in the Swiss Alps.
Purple mountain saxifrage below

Caucasus in Europe; the East African Highlands; the Himalayas in Asia; the Andes in South America; and the Rocky Mountains in North America. These regions vary enormously in terms of age, geographical location, rock types and wildlife, but all high mountains have one particular characteristic in common. They support distinct layers or zones of vegetation, between the lowlands and their snow-covered summits. The vegetation belts you encounter on the way up a high mountain reflect the changes seen as you move from the equator to either the northern or southern extremes of the globe. For example, a typical African mountain supports four zones. Montane forests are on the lower slopes; the bamboo zone starts at 2,500 metres (8,200 feet); the heath zone is found above 3,300 metres (11,000 feet); and then beyond this is the alpine zone, which lies immediately below the perpetual snows of the highest mountain peaks.

The higher you climb up a mountain, the colder the temperature becomes. As a world average, there is a temperature drop of 2.8°C for every 500 metres (3.1°F for every 1,000 feet) you rise above sea level. The mountains of the tropics have a slower decrease than mountains in temperate regions, but even on the equator there is a nightly frost above about 3,000 metres (about 10,000 feet). It is largely this fall in temperature, coupled with the amount of rainfall and the increasing winds, that determines which plants and animals live there.

Cold is the chief enemy of mountain wildlife. The great mountain summits, where temperatures sometimes fall to minus 20°C (minus 4°F), or less, are virtually without life. Lichens, which are not true plants but algae and fungi living together in partnership, are among the highest-living vegetation in the world – but even they cannot survive here. Many animals on the upper slopes escape the cold by moving to the lowlands in winter; other, smaller creatures cope by burrowing underground and hibernating. Some species, such as the North American mountain goat (Oreamnos americanus), have double coats which provide them with extra insulation. Many mountain-dwellers have short noses, ears and tails to leave fewer extremities open to heat loss.

Wind is another serious problem. On high mountains winds may blow at more than 160 kilometres (100 miles) per hour; speeds of 600 kilometres (375 miles) per hour have been recorded on at least one occasion. It is not surprising that typical mountain plants, such as the gentian, are not very tall and therefore cannot easily be blown away or damaged by the wind.

Another difficulty is thin air, which means less oxygen. Mountain animals cope by having more red blood cells than lowland species, giving them a very high ability to pick up and carry every scrap of available oxygen. Indeed, many 'alpine' animals and plants (those well suited to mountain life) are so highly adapted that they cannot survive anywhere else. And because the habitat on which they depend is so limited, mountain species are especially vulnerable to extinction.

The mountains themselves will be there for a long time to come, but mountain habitats are under threat. Though it may be hard to believe, the greatest danger comes from people. These days mountains attract enormous numbers of tourists who go to fish, hike, camp, ski, climb or just enjoy the clear air and glorious scenery. They put a surprising amount of pressure on mountain landscapes and wildlife.

But the most serious threat comes from mountain forest destruction. Most high mountains were once clothed continuously to the treeline in forest. The roots of the trees held the thin mountain soil in place, and together the trees acted as a sponge and allowed water to filter very slowly downhill. As land becomes scarce everywhere, people have been turning more and more to mountain forests for commercial timber, for fuel, food and building materials, or simply to clear more land for agriculture. Because the forests have gone, now when it rains the water rushes down the mountainsides in torrents. It destroys villages and agricultural land and takes the soils with it. In Nepal alone, over 200 million cubic metres (260 million cubic yards) of soil is washed from the Himalayas into the River Ganges and out into the Bay of Bengal every year. Tragically, it is a similar story in many parts of the world.

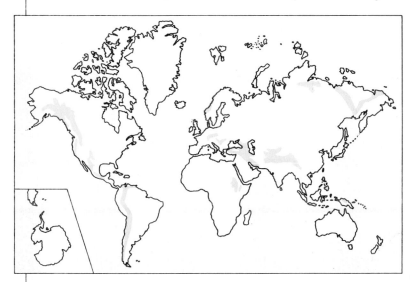

The world's mountain ranges
above
Mountains are found all over the world, from the Rockies to the Himalayas. The map shows the main ranges.

Mountain climbers

On the highest mountain peaks, the only life to be found is a few small creatures blown there by mistake and the occasional mountaineer who has ventured up of his own accord. But just below, on the steep upper slopes, there are small patches of hardy plants and a number of animals that are more sure-footed than any mountaineer could hope to be.

The chamois (*Rupicapra rupicapra*) is famed for its breathtaking leaps, which it makes with almost suicidal daring. Upward jumps of four metres (13 feet) and long jumps of over six metres (20 feet) are not unusual. The young animals learn their mountaineering skills very quickly, and after a few practice jumps on and off their mothers' backs they will happily follow the parents along narrow ledges or down precipitous screes. The chamois lives mainly in the Pyrenees, the Alps and the Apennines.

Chamois (*Rupicapra rupicapra*) *top right*
Klipspringer (*Oreotragus oreotragus*) *right*
Yak *below* On a mountain trail near Everest.

The African klipspringer (*Oreotragus oreotragus*) is perhaps an even greater master of the slopes. There have been claims that this antelope can land on a point no bigger than a large coin, and it seems to bounce up almost perpendicular cliffs like a rubber ball. Apart from its agility, the klipspringer has another form of protection against predators. Its coat is so brittle that, if caught, the hairs come away in the predator's mouth – and the klipspringer escapes. But unfortunately the mountain-climbing ability of many such animals, together with their shyness, has made them a challenging and highly sought-after quarry for sportsmen. The klipspringer's coat provides little protection against a rifle bullet.

The yak (*Bos mutus*), a rather strange mountain animal related to dairy cattle, also has an unusual coat. It is dense and matted for protection against the cold, giving it a rather awkward appearance. While it is not as agile as either the chamois or the klipspringer, the yak is still an expert climber and very sure-footed on even the most treacherous-looking slopes. Found in Tibet and the Szechwanese Alps in China, it is the highest-living wild animal in the world and often climbs to over 6,000 metres (19,500 feet) in its constant search for food. Somehow it manages to survive on seemingly bare slopes, by scraping a living on the sparse mosses and lichens.

Salamanders

The alpine salamander (*Salamandra atra*) is commonly found at altitudes of up to 3,000 metres (nearly 10,000 feet). Not only has it successfully adapted to cold mountainous conditions, like other alpine animals, but it has also

become accustomed to life without water. While most other salamanders and newts spend at least part of their lives in water, particularly during the breeding season, the alpine salamander is completely terrestrial. Even its young are born on land, and not as tadpoles but as miniature adults; at five centimetres (two inches) long they are a little under half the size of their parents.

The alpine salamander is also outstanding because it has the longest gestation (pregnancy) period of any terrestrial animal. In some parts of its range, in the alpine regions of Europe, individuals with pregnancies of over three years have been recorded. Generally the higher these amphibians live the longer the gestation period, but two years at heights of about 600 metres (2,000 feet) seems to be about the minimum.

Alpine salamanders are secretive animals and spend most of their time hiding under logs and stones or among low-growing vegetation. They usually only leave cover at night but will come out to shower during welcome rainstorms that take place in the daylight hours.

Alpine salamander (*Salamandra etra*) *left*

Dingoes

In the dense forests, open plains and mountainous areas of Australia lives a very close relative of the domestic dog: the dingo (*Canis familiaris dingo*). It was probably tame when it arrived in Australia along with people, some 8,000 years ago, and gradually became wild again. Dingoes are still kept – or captured as wild pups – by some aboriginals. Although occasionally trained as hunting dogs they are more often used as 'blankets' and taught to sleep in a huddle with their owners to provide protection from the cold. The number of dingoes in the huddle depends on how cold it is.

Four or five pups are born in the late winter or spring, in an underground burrow or cave. Although they become independent after three to four months they stay with their parents for as long as two years and provide them with help in raising the next litter.

Dingoes hunt mostly at night, either alone or in family groups, for rabbits, rats, lizards and birds. During periods of drought they may also take small kangaroos and some farm animals. The mistaken belief that these animals prey regularly on sheep has led to hundreds of thousands of dingoes being killed unnecessarily by sheep farmers over the years.

Dingo male *above*

ANDEAN CONDOR

DATA

SPECIES
Andean condor
(*Vultur gryphus*)

CLASSIFICATION
Falconiformes (eagles, hawks
and vultures)

DISTRIBUTION
The Andes and the Peruvian
coast

HABITAT
High mountain peaks and
coastlines

SIZE
Length of about 1.3 metres
(5½ft); wingspan up to 3 metres
(10ft)

FOOD
Living small mammals and
corpses of larger mammals;
along the coast eats seabirds and
their eggs, also dead fish, whales
and seals.

Anyone who has seen a condor in flight is unlikely to forget the spectacular combination of its enormous size and its ability to soar for hours without once needing to flap its wings. Measuring three metres (ten feet) from wingtip to wingtip, the Andean condor is the largest of the world's two condor species – though it was once dwarfed by an extinct relative, *Teratornis incredibilis*, which had a five-metre (16-foot) wingspan.

Andean condors, as their name suggests, live exclusively in the Andes. These members of the vulture family nest and roost as high as 4,000 metres (13,000 feet) up in the mountains and soar even higher, often sailing out over open savanna or occasionally as far as the Peruvian coast. Like all vultures they are beautiful and graceful in the air but hardly pretty close up. They have powerful, short, hooked bills for tearing flesh and naked heads and necks adapted for feeding inside carcasses without making their feathers filthy. In fact the adults often make considerable efforts to keep clean, wiping their heads and necks on grass or sand after feeding.

Andean condors feed almost exclusively on carrion, which is spotted from above or detected through the behaviour of other birds. These massive birds often go for several weeks without feeding, but when a suitable meal is found they eat up to two-thirds of their own weight at one sitting. Despite their size condors make no attempt to drive off eagles, coyotes or wolves already at the kill, preferring simply to wait patiently before feeding upon whatever is left. The birds are often so heavy after gorging that they are unable to take off on level ground and have to walk several hundred metres up a hill and run down before getting airborne.

Condors have an exceptionally slow breeding rate, which means that populations take a long time to recover from hunting pressures. They nest once every two years, lay only one egg and tend the chick closely for a year or more. It then takes at least a further six years for the young birds to develop sufficiently to breed themselves.

In the Andes, condors are killed for food by some highland Indians, who also use the wing bones for flutes, the quills for cigarette holders and eat the eyes to sharpen their own sight. Although their populations are small and widely scattered, the Andean condors' home is probably remote enough for them to survive. But in California it is another story; the California condor (*Vultur californianus*) is in imminent danger of extinction. There are fewer than 30 individuals left in the wild, all in one area, yet still each year a few birds are illegally shot and poisoned.

The chough (*Pyrrhocorax pyrrhocorax*) and alpine chough (*P. graculus*) are both members of the crow family. With their glossy all-black plumage they resemble crows, rooks, ravens and others in the group, but choughs (pronounced 'chuffs') differ in having colourful, slender, down-curved bills. These are ideally suited for their diet of worms, caterpillars, ants and beetles. Found in many parts of Europe and Asia, both species are among the record-holders for altitude attained by birds. One was actually seen to take off from a ledge 7,000 metres (23,000 feet) up on the slopes of Mount Everest. Choughs have a very buoyant flight and love to perform aerobatics, wheeling and soaring along the mountain and coastal cliffs on which they nest.

PTARMIGAN

The ptarmigan is well equipped for its life on high, barren mountain tops. It has excellent insulation against the cold, with downy body feathers, feathered legs and even feathered feet to use as snow-shoes. If the weather is particularly bad this bird simply sits in the snowdrifts waiting for conditions to improve, often going into a form of hibernation by reducing its heart and breathing rates to save energy. Only the severest mountain blizzards drive ptarmigans down to lower, more sheltered places.

While their wings and underparts remain white throughout the year, ptarmigans go through a complex series of moults to change the colour of their upper plumage from pure white in winter to mottled brown in summer. There are transitional stages in spring and autumn. This gives them excellent camouflage at any time of year.

These game birds are most visible in March and April, which are the peak months for courtship and establishing a territory at the breeding site. The males stand on boulders and other prominent display platforms, first to attract their mates and then to keep guard while the females incubate their eggs. Birds which fail to establish these territories are forced outside the main breeding area where they form small flocks. The eggs are laid in mid-May or early June in a shallow scrape on the ground, sometimes lined with a little grass or vegetation. The five to ten eggs hatch after three or four weeks and the young leave the nest almost immediately, though they cannot fly for a further ten days.

During the final stages of incubation, or after the eggs have hatched, male ptarmigans often form their own temporary flocks. But later they rejoin their family parties, which remain together until autumn before collecting with similar parties to make huge winter flocks of 100 or more birds.

Ptarmigans live on high mountains wherever there is bilberry, crowberry and heather – the shoots, leaves and fruits of these plants provide most of the birds' food. They occur in Scotland, Iceland, Greenland, Scandinavia, across northern Europe to the Bering Strait and across much of northern North America, where they are known as rock ptarmigans. The birds also occur in one or two mountain ranges further south, notably the Pyrenees and the Alps.

DATA	
SPECIES	Ptarmigan (*Lagopus mutus*)
CLASSIFICATION	Galliformes (game birds)
DISTRIBUTION	Arctic and sub-arctic regions of Europe and North America and several mountain ranges further south
HABITAT	Mountain tops and tundra well above the treeline
SIZE	Head-to-tail length 35cm (14in)
FOOD	Shoots, leaves and fruit of bilberry, crowberry and heather

GIANT PANDA

DATA

ENDANGERED

SPECIES
Giant panda
(*Ailuropoda melanoleuca*)

CLASSIFICATION
Carnivora (carnivore)

DISTRIBUTION
Szechwan, Kansu and Chensi
provinces in China

HABITAT
Mountain forests and dense
stands of bamboo

SIZE
Head and body length up to
1.5 metres (5ft) plus tail of
about 13cm (5in); weight up to
160kg (350lb)

FOOD
Mainly bamboo shoots; some
other plants and even small
animals occasionally taken

According to an ancient Chinese legend the giant panda got its markings from grief. A young girl once grew up with a group of pandas, which in those days were all pure white, near the mountain of Wolong. One day a leopard attacked one of the baby pandas and the girl defended the young animal and saved its life, but later she died from her own wounds. At the funeral all the pandas of the country were present to pay her tribute, each wearing black on their shoulders, arms and legs in mourning. As they wept they rubbed their eyes with their front paws and created black eye patches. They put their heads in their paws, making a black nose, then reached up and stained their ears. According to the legend, all pandas since have worn these distinctive markings.

Today it is the pandas themselves which are dying. Once widespread in China, they now occur only within a restricted range in the mountains of the south-west, where between 400 and 1,000 still survive.

Their problem is caused by an unhappy combination of human activities and a freak of nature. The two species of bamboo on which they rely for food follow a hundred-year cycle, so that roughly once every century the plants bloom, drop their seeds and die. In ancient times, when bamboo died in one area, the pandas were able to move into another. Today people have taken over more and more land for agriculture and this is no longer possible. Nowadays, when the bamboo dies the pandas have nowhere else to go, so they starve.

Although scientists believe that widespread flowering of bamboo in panda country is imminent, dozens of animals have died and the impending danger has rallied conservationists all over the world, led by the World Wildlife Fund (the panda has been its symbol since 1961) and the Chinese government in a massive 'Save the Panda' campaign.

Pandas are very elusive animals and despite years of work on almost 50 captive individuals in zoos all over the world, surprisingly little is known about them. Research into their requirements is clearly essential to their long-term survival, but in the meantime everything possible is being done to save the species from extinction. Extra food has been supplied and a network of panda reserves established. It is an enormous task, and there is very little time, but anything that may help is worth trying.

JAPANESE MACAQUE

High in the mountains of Japan lives the most northerly and one of the most interesting monkeys in the world. The Japanese macaque, or snow monkey, illustrates perfectly how monkeys are capable of learning useful patterns of behaviour from one another.

One winter, a number of years ago, scientists were observing a macaque troop living in the chilly mountains of northern Japan. The animals started to explore a part of the forest in which they had never been before. Although macaques can climb trees (and usually sleep in them) they spend most of their time on the ground, and a few of the animals soon came across a hot volcanic spring. Since the air was freezing cold and the snow was over a metre deep, they tried bathing in the spring to keep warm. Soon the habit spread and now all Japanese macaques in the area take hot baths every winter. Whole troops can often be seen with only their heads sticking out of the water – sometimes covered with several centimetres of snow.

Unfortunately the hot springs do not provide food so the macaques eventually have to get out, soaking wet, and brave the snowstorms and blizzards to find something to eat. Their long, shaggy coats protect them from the cold while they feed on fruit, leaves, insects and other small animals.

Some individuals of this species have also learnt to wash their food in water. Those near the coast often dip items in the sea, apparently because they like the salty taste. Some have even taken to carrying sweet potatoes and other delicacies in their arms while they walk like humans on two legs, down to the water's edge. There are sometimes as many as 700 Japanese macaques living together and these 'tricks' spread very quickly within the troop.

Although very large troops occasionally occur, these monkeys normally live in social groups of about 150 to 200 individuals. In fact troop size used to be lower, but when Japanese researchers started to feed the animals in order to study their behaviour, the average number in a troop rose.

In most troops, females predominate, with several generations of daughters living with their mothers throughout their lives. The young are born singly after a gestation period of six to seven months. The troop is led by a dominant male; other males are forced to leave when they become adolescents and they move off to join

drifting, peripheral bands of other young males. Eventually a mature male will join a troop to which he is unrelated.

Snow monkeys have always played a special role in Japanese mythology, folklore and art. They are used to represent the wisdom of Buddha: 'see no evil, hear no evil, speak no evil' is symbolized as one covers his eyes, the second blocks his ears and the third covers his mouth. Although traditionally protected in Japan, their numbers have been declining in recent years due to habitat destruction.

Other macaques of note are the Barbary ape, the pig-tailed monkey, and the Rhesus monkey, famous as a subject of medical research – particularly into blood groupings. As a group the macaques are active by day and have excellent sight, hearing and smell.

DATA
SPECIES
Japanese macaque (*Macaca fuscata*)
CLASSIFICATION
Primates (apes, monkeys, lemurs and others)
DISTRIBUTION
Japan
HABITAT
Thick forests of steep mountain slopes
SIZE
Head and body length up to 60cm (2ft); tail a further 10cm (4in)
FOOD
Fruit, leaves, insects and other small animals

SNOW LEOPARD

DATA

ENDANGERED

SPECIES
Snow leopard (*Panthera uncia*)

CLASSIFICATION
Carnivora (carnivore)

DISTRIBUTION
Himalayas and some other high mountain ranges in central Asia

HABITAT
High-altitude coniferous forest, scrubland and mountain steppe

SIZE
Head and body length up to 1.5 metres (5ft); tail length 90cm (3ft)

FOOD
Ibex, wild sheep, marmots, mice and other small animals, and birds.

In the Himalayas and some of the other high mountain ranges in central Asia lives one of the most endangered cats in the world, the snow leopard. Rarely seen in the wild, it is a shy, mostly nocturnal animal that nowadays avoids people by living at high altitudes in difficult and unwelcoming terrain. Enormous numbers of snow leopards have been killed in the past, by farmers concerned about them taking domestic livestock but also for sport and because their fur is very valuable. Such hunting pressures have caused a dramatic decline in their population in recent years. Today there are probably only a few hundred left.

Snow leopards, or ounces as they are sometimes called, are perfectly adapted to life in the high mountains. They grow specially thick coats in winter and the soles of leopard feet are covered with cushions of fur to help them walk over the snow and to protect their paws from the cold. The tails are also covered with thick fur, to retain heat, and are very long to act as a counterpoise when balancing. This is important when the leopards are climbing trees, moving over steep ground or springing on prey.

These animals are ideally camouflaged for their environment. The fur is pale grey with the characteristic leopard spots, which are in fact rosette-shaped black areas when seen in close-up. These patches break up the leopard's body's outline by mimicking the small patches of shadow and shade seen in the snowfields and glaciers of their habitat. All big cats depend on camouflage in order to catch their prey, since the latter are usually fast and agile and so must be taken by stealth and then a surprise dash or 'charge'. The snow leopard has short but muscular and well-developed legs for brief bursts of speed after the prey has been stalked to within charging distance.

There is some dispute among scientists as to whether the snow leopard should be classified in the same genus, *Panthera*, as the four biggest cats – the leopard, jaguar, tiger and lion. Some experts put it in a genus of its own so that its Latin name is *Uncia uncia*.

These big cats feed on ibex, wild sheep, marmots, mice and birds. Larger animals are occasionally captured in ambushes, but normally they are stalked and sprung upon from up to 15 metres (nearly 50 feet) away. Snow leopards tend to follow their prey, which migrate up and down the mountain slopes during the seasons to avoid the worst of the snow. In summer they may live as high as 6,000 metres (19,500 feet) but in winter move, with their prey, to 1,800 metres (6,000 feet) or lower.

Although much of their lives is spent alone, snow leopards come together early in the year to mate. The cubs are born in April, May or June, after a gestation period of about three months. They spend their first weeks safely hidden in a rocky shelter lined with their mother's fur. Gradually the cubs venture outside to follow her and learn the necessary hunting skills. Together they roam over her territory, which may be as large as 100 square kilometres (40 square miles), covering the entire area roughly once every week until well after the cubs' first winter. Such enormous ranges are necessary since the cold, barren land yields little prey.

MOUNTAIN LION

The mountain lion is known by a variety of names, including puma, cougar and panther. It is the largest and probably the best known of all North American cats. It certainly makes its presence obvious with a range of calls, many like those of a domestic cat but louder, and a piercing scream which has a completely unknown function.

Although found in many different habitats, including forests, grasslands and deserts, mountain lions prefer rocky and mountainous terrain. They are extremely agile and may leap up to a ledge or tree branch more than five metres (16 feet) above the ground. These big cats often hunt over enormous areas and may take a week or more to complete a single circuit of their range, which in summer extends up to 300 square kilometres (115 square miles) though in the winter it tends to be less than half this size. Very often the ranges of individuals overlap, but neighbours seldom use the same localities at the same time in order to avoid unnecessary fighting.

Most births occur late in the winter or in early spring. Three to four kittens, which have spotted coats until they are about six months old, are born in dense vegetation, a rocky crevice or a cave. They usually venture out with their mother by autumn and are making their own kills by the end of winter, though they remain with her for a year or more before dispersing to find homes of their own.

Deer are the lions' most consistent food, which they stalk and seize after a swift dash. They often eat rodents and rabbits as well. Most hunting takes place at twilight and the larger kills are usually dragged to a sheltered spot to be eaten. Leftovers are hidden under leaves and debris, ready to be visited for additional meals over the next few days.

Mountain lions themselves have been hunted intensively in the past – and still are in some places – by people who view them as a threat to domestic animals such as horses and sheep. Dogs pursue the lions until they seek refuge in a tree, where they are easily shot. Although still widespread, they have been eliminated in many areas, and two subspecies, the Eastern cougar and the Florida cougar, are now threatened with extinction.

DATA

SPECIES
Mountain lion
(*Felis concolor*)

CLASSIFICATION
Carnivora (carnivore)

DISTRIBUTION
South-western Canada, through parts of USA into Central and South America

HABITAT
Wide range, from rugged mountains to tropical forests and deserts

SIZE
Head and body length up to 2 metres (6½ft); tail up to 78cm (30in)

FOOD
Mice, rats, hares, deer and other mammals

SWAMPS AND MARSHES

Brazillian swamp *below*
On the Rio Grande in Do Sul State, South Brazil, broad stretches of swamp are interrupted only by grasses.
Cypress swamp *right*
This bald cypress swamp is in Illinois, USA.

Mountain bogs, lowland estuaries, mudflats, tropical swamps, salt marshes and fenland are just some of the world's many varied swamp and marshland habitats. To humans, few of them are particularly attractive places – they are often unsheltered, soggy and muddy, or buzzing with mosquitoes. People the world over tend to regard them as wastelands. So, perhaps not surprisingly, these 'wetlands' are under greater threat than almost any other habitat in the world.

City planners convert swamps into rubbish tips; sewage from urban areas and chemical wastes from factories are dumped on marshes; they are favourite sites for nuclear power stations and oil refineries; dams are built upstream from them, for hydroelectric schemes or water storage; farmers 'improve' these wetlands by draining away the water to provide more land for agriculture; some are drained in the hope of reducing the numbers of disease-carrying insects, such as mosquitoes; and they are often the final resting places for enormous concentrations of pesticides and chemical fertilizers which enter via rivers from nearby farmland.

Sadly, such large-scale pressures and destruction are not restricted to developed countries. In many parts of Africa, South America and South-East Asia wetlands are being subjected to similar fates and many are likely to be damaged or destroyed in the near future.

As the name implies, wetlands are areas which in some way contain both land and water. For example, many African and Australian 'wetlands' are often dry for years at a time until the rare rains return; while swampy regions may be a permanent mixture of islands and shallow lakes; and other areas may be flooded for most of the time except for the occasional drought.

Together the wetlands occupy a very small proportion

– just a few per cent – of the planet's surface. Yet swamps and marshes provide ideal conditions for wildlife. Exposed to and warmed by the sun, swamps and marshes act as traps for nutrients and organic matter. This makes them exceptionally productive, harbouring enormous growths of microscopic plants, which in turn feed microscopic animals at the beginning of many rich food chains. Indeed, some wetlands can produce more than 50 times as much plant material as a similar area of natural grassland, or eight times as much as is produced in a cultivated field. Even the water run off from swamps and marshes is very rich, fertilizing the lakes and seas downstream.

Many wetlands are vital to the survival of enormous numbers of wildlife which are in turn of great importance to man. For example shrimps, the world's most valuable wild animals in terms of food harvest, are born at sea but move into wetlands for food and shelter. The shrimps' close rivals are cod and herring, both of which also spawn at sea but the young hatchlings migrate to estuaries, to feed and grow during the first two years of their lives.

Overall, no less than two-thirds of commercially important fish and shellfish harvested along the Atlantic seaboard of North America depend on estuaries and associated wetlands for food, spawning grounds and nurseries for their fry. Europe's most famous wetland, the Wadden Sea, which lies along the Dutch, German and Danish coasts, is equally important. Comprising 8,000 square kilometres (3,100 square miles) of shallows, tidal flats and salt marshes, it supports nearly 60 per cent of the North Sea's population of brown shrimps; over 50 per cent of its sole; 80 per cent of its plaice; and almost 100 per cent of its herring population at some stage in its life.

Many swamps and marshes are also vital to the survival of birds such as waders, ducks and geese, which use them as resting stops on migration. Enormous numbers of other species including grebes, herons, moorhens, kingfishers, reed warblers and yellow wagtails are also present for much of the year. The Ouse Washes, one of Britain's few remaining wetlands of international importance, constitute the largest inland gathering ground for wildfowl in the country. This vast ribbon of flat and seasonally flooded land, some 22 kilometres by one kilometre (14 miles by half a mile) is winter home to one in five of all north-west Europe's Bewick's swans, as well as thousands of pintails, wigeons, pochards, mallards and other birds.

Among the many other animals and plants specially adapted to, and relying upon, wetlands are the extra-ordinary lungfish of South America, Australia and Africa. Two of the three species can survive a complete lack of oxygen in the water by breathing air, and the third can live without water entirely. The Australian lungfish burrows into the mud while it is still damp and envelops itself in a mucous cocoon. This dries to form a protective case and the fish goes into a kind of hibernation, living off its reserves of fat with just enough air filtering through the top of the case to breathe, until the rains return.

Swamps and marshes have many other attributes, as well as providing havens for wildlife. They act as natural water-purification systems, removing silt and filtering out or absorbing many pollutants; those on the coast help to blunt the force of major storms; others even act like sponges, mopping up water when there is too much coming down a river and releasing it gradually later, thereby preventing catastrophic flooding and massive erosion downstream.

The vanishing wetlands

But wetland destruction is continuing as fast as ever. Governments often believe that there are powerful economic arguments for siting huge industrial developments on coastal wetlands: nuclear power stations need a large and regular supply of cooling water, while oil refineries are more efficient if sited near deep water in estuaries, where supertankers can dock alongside. These days there are also tremendous pressures on otherwise 'unused' land, to meet ever-increasing demands for food and to provide room for housing.

Too often, though, politicians and planners forget to look at the other side of the equation. Does the detrimental impact on the environment outweigh the value of such developments? For example, what is the point of draining a marshland for crops if it destroys a fishery of far greater importance to the local or national economy, or ruins a wildlife community or even a thriving tourist industry?

Fortunately, there are some hopeful signs for the future. Some governments were sufficiently interested in wetland conservation to form the Ramsar Convention (otherwise known as the Convention on Wetlands of International Importance) which came into force in 1975. This is an agreement in which member countries must not only promise to promote wetland conservation in general, but must actually designate sites of international importance within their own borders and agree to conserve them. There seems at last to be a dawning realization that swamps and marshes are not wastelands fit only for 'improvement' but in fact play important roles in all our lives.

Wetland birds

Many of the world's 8,800 species of birds live on or around swamps and marshes. As well as the familiar ducks, geese and swans, there are also huge breeding populations of herons, storks, ibises, cormorants, spoonbills, flamingos and waders, various birds of prey and myriad smaller birds in these wetland habitats.

In most parts of the world the heron family is among the best-known birds associated with wetlands. There are 60 species, distributed worldwide except for the extreme north, the Sahara and Arabian deserts and some oceanic islands. They are divided into four main groups: bitterns, tiger herons, day herons and night herons. As their name suggests the night herons are mostly nocturnal but also feed by day during the breeding season and at other times.

Another well-known wetland bird is the kingfisher. There are nearly 90 species of kingfishers around the world, most of them living near water and feeding on small aquatic animals such as fish and dragonfly larvae. The malachite kingfisher (*Alcedo cristata*), for example, can often be seen perched on the vegetation fringing lakes, swamps and marshes. Found throughout Africa south of the Sahara, it is one of the most common birds on the entire continent.

Night heron (*Nycticorax nycticorax*) top
American jacana (*Jacana spinosa*) above

The strikingly coloured jacanas are another common group of waterbirds. The African jacana (*Actophilornis africana*) is well known for its habit of hiding underwater from predators, leaving only its bill and nostrils above the surface. As a group, jacanas are distinguished by their extremely long toes and claws, which enable them to walk easily on floating vegetation and the leaves of waterlilies, hence their name of 'lily-trotters'.

Jacanas are widespread throughout the marshlands, rice fields and freshwater shores of the tropics. All seven species feed on a mixed diet of aquatic insects, snails and occasionally small fish and the seeds of aquatic plants. Breeding takes place at the beginning of the wet season, when insect food is most abundant. The male is unusual in the avian world since he performs all the duties of nest building, incubation and care of chicks.

Malachite kingfisher
(*Corythornis cristata*) above

Dragonflies

'Dragonfly' is a fitting name for these carnivorous terrors of the insect world. They are powerful fliers, some species being able to reach speeds of up to 50 kilometres (30 miles) per hour and seize prey in flight as they go. Even their nymphs (larvae), which live underwater and breathe with special gills, eat any living thing that they can catch and hold. This includes tadpoles, small fish and even other dragonfly nymphs.

After months or years in the water as immature nymphs, dragonflies leave the water for good when they become adult. They are normally seen near fresh water, though some species do migrate over land and sea. The nymph crawls up the stem of a plant standing in the water and, from a split in the skin on its back, the adult slowly and deliberately squeezes into the air.

Their fearsome habits and appearance hide the fact that dragonflies are very beneficial to man, as great destroyers of pest insects. They were once widely feared and given names like 'horse-stinger' and 'devil's darning needle', which dates back to an old superstition that they could sew up children's mouths. These fears are of course unfounded.

There are over 5,000 species of dragonflies and their relatives, the damselflies. Their fossil relatives were among the first groups of insects to fly and some grew to huge sizes, with wingspans of over half a metre (20 inches) in some cases. Apart from their size dragonflies have remained much the same for 300 million years, though they have adapted throughout the ages to exploit the insects of their day.

Damselflies (Pyrrhosoma nymphlua) left The female oviposits in the water while the male clasps her behind the head.

Frogs

There are over 2,600 species of frogs and toads, found in most parts of the world where there is fresh water or moisture. A few hardy species live north of the Arctic Circle, in water barely above freezing. Some are adapted to life in trees, others burrow underground and a few even live in deserts. But they all depend on water because they all lay eggs known as spawn which can develop only in water or damp places. This usually means that, for the first part of their lives at least, frogs and toads are aquatic – though many go ashore later on.

To cope with this dual existence, these amphibians have three different breathing systems. The young tadpoles breathe through gills, which take oxygen out of the water. As their legs develop, the gills and tail are lost and the tadpoles become small frogs that breathe air with lungs. The third system is also found in adults, whose skin is richly supplied with blood vessels and so can absorb oxygen directly into the blood from the air and the water.

Toads tend to have dry, warty skins and move around by crawling, while frogs have smooth, moist skins and travel by jumping. Indeed frogs are famous for their leaping, which is a way of getting from one place to another as well as a very effective method of escaping from an enemy as quickly as possible.

Green frog (Rana clamitans) above

95

AMERICAN ALLIGATOR

DATA

SPECIES
American alligator
(*Alligator mississippiensis*)

CLASSIFICATION
Crocodilia (crocodiles and
alligators)

DISTRIBUTION
South-eastern USA

HABITAT
Marsh-bordered lakes,
freshwater and brackish coastal
marshes

SIZE
Up to 6 metres (20ft), though
usually considerably less

FOOD
Young feed on insects, worms,
frogs and small fish; medium-
sized individuals prefer fish and
turtles; adults take birds and
small mammals

The American alligator was once a very common animal in large marsh-bordered lakes, coastal marshes and other wetland habitats in south-eastern USA. But by the middle of this century hunting, persecution and habitat loss had forced it almost to extinction. Fortunately, severe protective measures were introduced in time and the population has since recovered significantly. In Louisiana, for example, there were around 26,000 alligators in 1957; today there are over 300,000.

Alligators are now found from the coastal parts of North Carolina westward to the southern portions of South Carolina, down into Georgia, Alabama and Florida, through much of Mississippi and across Louisiana into eastern Texas. The total population now stands at around 800,000 but, while the species is not seriously endangered, it is important that harvesting and persecution are strictly monitored to prevent a repeat of the past near-disaster.

The American alligator is one of seven species of alligator, belonging to a group of reptiles which also includes the gharial and 13 species of crocodile. The group contains the closest living relatives of the dinosaurs. There are a number of differences between alligators and crocodiles, but the main one concerns their teeth. An alligator has 17 to 22 teeth on each side of its jaw, while a crocodile has 14 to 15; also, in an alligator the fourth tooth on each side of the lower jaw fits into a pit in the upper jaw and so is invisible when the mouth is closed, whereas in a crocodile there is no such pit and these teeth are therefore always visible.

All crocodiles and alligators lay white, hard-shelled eggs roughly the size of a hen's egg. Around 20 to 30 are laid in nests dug by the females at the water's edge. The females remain nearby to guard them, often lying in the sun or floating underwater with just their noses, eyes and ears showing above the surface. The eggs are incubated by the heat of decaying vegetation placed inside the nests. The young animals break out of their shells with the aid of a special egg-tooth on their snouts, which is shed soon after hatching. The hatchlings grow rapidly, from about 25 centimetres (ten inches) when they are first born, to twice this length by the end of their first year, and twice as long again by the age of two. They continue growing for many years, and in the case of species such as the American alligator may reach an imposing six metres (20 feet) or more.

ANACONDA

The anaconda is the largest reptile in the world, though it has been the subject of more exaggerated claims about its size than almost any other living animal. There are many 'sightings' of individuals 15 metres (nearly 50 feet) or more in length, though in reality nine metres (30 feet) is probably the maximum. Even so, in combination with their massive bodies this makes them considerably larger than any other species of snake, crocodile or turtle.

Early Spanish settlers in South America named the anaconda 'matatoro', which means bull-killer. Although fully grown bulls are unlikely to feature in their diets, in theory anacondas do grow large enough to eat bullocks – and even people. But there have been few verified accounts of this happening. They do occasionally bite if frightened, and this can be very painful, though the bite itself is not dangerous. However, even the largest specimens would have difficulty eating a whole person.

Anacondas are exclusively aquatic, though their normal prey consists mostly of land mammals and birds that come to the water's edge to drink. These are killed by suffocation as the victim is crushed to stop it breathing; its bones are rarely broken, and it is swallowed whole.

Anacondas belong to a group of snakes known as the boids, which includes all the boas, pythons and other large snakes alive today. They are virtually silent animals – at most they can utter a hiss – and are entirely deaf. However, they have a good sense of smell, can accurately detect movements by vibrations in the ground on which they lie, and have fairly good eyesight, enabling them to distinguish outlines well at short distances and to recognize immobile objects.

Very agile in the water, anacondas are constantly diving and surfacing and in the process often cover considerable distances. They have been known to undertake extensive trips into the open sea, sometimes floating along with drifting tree stumps for long distances. They climb as well as they swim and often use trees along the edge of a river or marsh in which to take refuge.

DATA

SPECIES
Anaconda (*Eunectes murinus*)

CLASSIFICATION
Squamata (snakes and lizards)

DISTRIBUTION
Northern South America, primarily in and near the Orinoco, Amazon and the old Guianas

HABITAT
Rivers and swamps

SIZE
Lengths of more than 9 metres (30ft) and girths of over 1 metre (3ft) have been recorded

FOOD
Young feed on rodents, birds and other small animals; adults take mostly small deer, monkeys, goats and wild boars, but also waterside birds and even small cayman

MUDSKIPPER

DATA

SPECIES	
Mudskipper (*Periophthalmus* sp.)	
CLASSIFICATION	
Perciformes (perch, butterfly-fish, sunfish and others)	
DISTRIBUTION	
Tropical coasts of Old World	
HABITAT	
Mangrove and other coastal swamps	
SIZE	
Up to 20cm (8in)	
FOOD	
Insects and small shellfish	

Three hundred and fifty million years ago, in a mangrove swamp somewhere on our prehistoric Earth, a few fish began to haul themselves out of the water. They found a rich food supply of crustaceans and other small animals in the mangrove mud and soon made a regular habit of leaving the water to feed. Gradually overcoming the problems of moving on land and breathing air, these fish became the world's first back-boned creatures to colonize the land. A group of fish alive today, known as the mudskippers, are also able to live partly in water and partly out of it. Although not closely related to those enter-prising creatures 350 million years ago, they have developed very similar adaptations and behave more like newts than fish.

Ranging from a few centimetres to over 20 centimetres (an inch or so to nine inches) in length, mudskippers live in the mangrove swamps and muddy estuaries of many parts of the tropics. They spend most of their time out of the water, feeding on insects, tiny worms, small crustaceans, algae and other plants (depending on the species) in the mud. They even carry out their fights against rival males and courtship displays for females well away from the water's edge. Some species actually build low mud ridges, several metres long, around their terri-torial boundaries to keep out unwelcome neigh-bouring mudskippers. Others make a habit of climbing up the trunks of nearby mangroves.

Mudskippers have two main ways of moving on land. They can move very rapidly, in a series of little jumps, by curling their tails sideways and flicking them out straight. Or they can move slowly, and more precisely, using their front fins as crutches. They breathe in much the same way as crabs do on land: by keeping their gill chambers and mouths full of water, returning regularly to the water's edge to collect new, freshly oxygenated mouthfuls. But, unlike crabs, they are also able to take in oxygen through their skin, like a frog – so long as the skin is kept moist, which they do by rolling on their sides.

MEDICINAL LEECH

There are more than 300 species of leech living in the sea, in fresh water and even on land. They all have suckers at both ends, with which they cling on to plants or animals to feed on their sap or blood. The medicinal leech can 'drink' up to five times its own weight of blood at one sitting, leaving behind a Y-shaped cut in the prey's skin

caused by its strong teeth. But digestion is slow and the leech may require as long as seven months to absorb a single meal. Some individuals have been reported to go without food for 18 months.

Like their relatives the earthworms, leeches are hermaphrodite animals. Each individual has both male and female sex organs but cannot fertilize itself. When two leeches come together to mate, each exchanges sperm with the other so that they both become 'pregnant'. Eggs are then laid in a slimy cocoon which is secreted by a specialized row of segments towards the middle of the body known as the clitellum or 'saddle'. The clitellum is normally insignificant but becomes swollen and visible during the breeding season.

Medicinal leeches live under stones in freshwater ponds, streams and marshes. The young feed on frogs and tadpoles while the adults gorge themselves on blood, selecting mostly horses, cattle and other large mammals which come to the water to drink. But the demise of the working horse, the advent of piped water supplies for livestock and the drainage of ponds have all contributed to the decline of this species in most parts of western and southern Europe. The greatest threat to its survival, however, has come from over-collecting for the medical profession. When it feeds, the leech produces an anticoagulant chemical which prevents blood from clotting and so allows it to flow freely. Medicinal leeches were once widely used for blood-letting and in the 18th and 19th centuries. There was such a demand for the animals that they became extinct in many countries.

There is a renewed interest in medicinal leeches today. The anticoagulant they produce, called hirudin, is being used to research the mechanism of the human blood-clotting process. Hirudin is now supplied on a commercial basis but it is estimated that 12,000 kilograms (26,500 pounds) of leeches are used for its production every year, since thousands of the animals are required to obtain minute quantities of purified hirudin.

In the future it is hoped that hirudin and other useful substances available from medicinal leeches can be produced artificially. But while large-scale collection from the wild could stop by the end of the century, another threat is gradually taking over – the widespread loss of the leeches' marshland habitat.

DATA
ENDANGERED
SPECIES
Medicinal leech (*Hirudo medicinalis*)
CLASSIFICATION
Annelida (earthworms, rag-worms and leeches)
DISTRIBUTION
Parts of western and southern Europe
HABITAT
Under stones in marshes, ponds and streams
SIZE
Up to 13cm (5in) long and 1.5cm (½in) wide
FOOD
Young feed on frogs and tadpoles; adults suck blood of warm-blooded animals

CAPYBARA

DATA

SPECIES
Capybara (*Hydrochaerus hydrochaeris*)

CLASSIFICATION
Rodentia (squirrels, rats, mice and others)

DISTRIBUTION
Panama and South America east of the Andes

HABITAT
Dense vegetation around ponds, lakes, streams, rivers, marshes and swamps

SIZE
Shoulder height up to 60cm (2ft); weight up to 70kg (154lb)

FOOD
Aquatic plants, mainly short grasses

The capybara is the world's largest rodent. It looks rather like its close relative the guinea-pig, but is many times the size. Some extinct capybaras were even bigger than those alive today – probably twice as long and up to eight times as heavy.

Capybaras live in the dense vegetation around water and marshes in Panama and South America east of the Andes. Semi-aquatic animals, spending a great deal of time on land but often standing in water up to their stomachs, capybaras feed on plants. They also use water as a place of refuge and run into it as fast as they can when alarmed. They swim and dive easily and have even been known to hide in floating vegetation, with only their noses exposed above the surface to breathe. In an emergency a capybara is able to stay underwater for as long as five minutes. These rodents also spend a great deal of time wallowing in water or mud, to keep cool and to prevent their skins from drying out in the hot midday sun.

Capybaras are fairly nervous animals, and although active in the morning and evening in undisturbed areas they have become nocturnal in other parts of their range. They are considered to be agricultural pests in some places and are

hunted intensively for their meat and hides. The thick, fatty skin also provides a grease which is used in the pharmaceutical trade, and the incisor teeth are used as ornaments by local people. However, although capybaras have disappeared from some areas they remain widespread and common and are even raised commercially on special ranches.

Capybaras are usually found in groups of about 20 individuals, though there are reports of colonies of 100 or more. Each group has a very stable and viciously enforced social hierarchy in which certain individuals take preference over others during feeding, breeding and other activities. They are conspicuous animals, always fighting and arguing, and they make a range of different noises: a clicking sound to show contentment; a kind of purring to keep the group together; long, piercing whistles; and a variety of grunts and barks used mostly as alarm signals.

Unlike many of their relatives, young capybaras are born well developed with all their hair and can see almost immediately after birth. They are able to follow their mothers and eat within a very short time, but they tire quickly and are vulnerable to predators such as dogs, cayman (small crocodiles), foxes, and jaguars.

INDIAN RHINO

Like all rhinoceroses, the Indian rhino has suffered tremendously in recent years, from ruthless hunting for its horn. Enormous numbers have been killed by poachers and the species has disappeared from many parts of its former range. There are now fewer than 1,500 left in the wild.

Many people believe that rhino horn has special medicinal properties. It is used as an aphrodisiac in parts of northern India and elsewhere, while in China and neighbouring countries in the Far East it is popular for treating fevers, headaches, heart and liver problems and skin diseases. But any successes in these treatments are purely psychological because rhino horn is actually made of keratin, a hair-like substance which also forms the basis of hooves and fingernails. As far as we know it has no medicinal properties at all.

Also contrary to popular belief, in the wild the Indian rhino does not use its horn to attack enemies. It charges with its mouth open, using its sharp-pointed lower tusks instead. Although this rhino looks an awkward animal, especially as its folded skin gives the appearance of armour plating, it is suprisingly agile and may charge at speeds of up to 50 kilometres (30 miles) per hour. Indian rhinos occasionally attack elephants which are being used by people as observation platforms in the sanctuaries where they occur, but normally they only charge when wounded or when they feel their calves are threatened. Most individuals seek to escape, rather than attack, an enemy. Either way they are unfortunately ill-equipped to cope with modern man armed with sophisticated weapons.

Indian rhinos are mostly solitary animals. They spend the majority of their time near water, bathing daily and often wallowing in mud for hours on end. When the mud dries it forms a crust that helps to protect the animals from the sun and biting insects. Although active by day and during the night they feed mostly in the morning and evening. Enormous tunnel-like paths through grass that can grow eight metres (25 feet) high lead them to their grazing grounds, where they feed on grass, reeds and twigs.

Dung heaps are often found near the entrances to these tunnels or close to a favourite wallow. An individual will shuffle through the communal heap and contribute to it. The rhinos also have scent glands on their forefeet and leave scent trails. The trails form an important part of the communication system between these territorial animals, who have poor vision but a keen sense of smell.

The single young is born any time from the end of February through to the end of April, after a 16-month pregnancy. At birth the calf weighs up to 75 kilograms (165 pounds) and can have a shoulder height of 60 centimetres (two feet). As they get older the males grow considerably larger than the females. Indian rhinos are believed to have a lifespan of 50 years or more.

DATA
ENDANGERED
SPECIES Indian rhino (*Rhinocerus unicornis*)
CLASSIFICATION Perissodactyla (tapirs, rhinos, horses and zebras)
DISTRIBUTION Assam, west Bengal and Nepal
HABITAT Tall grass and reedbeds in swampy jungles
SIZE Up to 1.9 metres (6ft) at the shoulder; weight 2.2 tonnes or more
FOOD Grass, reeds and twigs

PROBOSCIS MONKEY

ENDANGERED

The Proboscis monkey (*Nasalis larvatus*) is the elephant of the monkey world. It has an enormous nose, sometimes over seven centimetres (three inches) long, which hangs down over the mouth and occasionally even touches the chin. The nose is used as a loudspeaker for warning honks, which makes it sound like noisy geese rather than a monkey when heard from a distance. Males have the longest noses and when the monkeys get angry or excited these swell and become red, in much the same way as people blush when they are embarrassed.

Proboscis monkeys live around the mangrove swamps and riverbanks of Borneo, in South-East Asia.

LAKES AND RIVERS

No less than 97 per cent of our planet's water is salty. But the remaining three per cent, which includes mountain streams, village ponds, man-made reservoirs and gravel pits, enormous rivers, canals and massive lakes, provides a wide variety of habitats and is occupied by a bewilderingly diverse selection of specialized plants and animals.

Freshwater habitats vary enormously in terms of age, size, depth and water movements, as well as through the climate and local geography of the areas in which they are found. Among the millions of kilometres of rivers throughout the world, the longest stretches are the Amazon in South America and the Nile in Africa. They are both over 6,400 kilometres (4,000 miles) in length, and the Amazon is so enormous that at any one time two-thirds of the world's fresh water is flowing between its banks. When in full flood it discharges 200,000 cubic metres (260,000 cubic yards) of water into the Atlantic Ocean every second. In complete contrast the world's shortest river is found in the USA and is only 134 metres (440 feet) long.

There is an equally dramatic range in the size of lakes, which may be parts of rivers or their birthplaces. Compare a tiny village pond with North America's Lake Superior, which is the largest freshwater lake in the world. It covers an area of 82,350 square kilometres (31,810 square miles). The village pond may be considerably less than one metre (yard) deep – but Siberia's Lake Baikal, the world's deepest lake, is more than one and a half kilometres (almost one mile) from the water's surface to its bed.

Strictly speaking, the only difference between a lake and a pond is the depth of the water and not the area or volume. In a pond the water is shallow and plants are usually able to grow anywhere on the bottom. But in a lake the bed falls away to greater depths where no rooted plants can grow because not enough light penetrates, and the water is cold and dark. Many lakes are so deep that we are still not sure what, if anything, lives near the bottom. In theory it is quite possible that many unknown and perhaps prehistoric creatures, such as Scotland's Loch Ness Monster, live in these deep waters where they are rarely, if ever, seen.

The shores of lakes and ponds provide most shelter, light and food and this is where most lake-living animals and plants live. The good swimmers and fliers, such as birds and fish, tend to be the ones which make their homes out in the middle, in the open water. Ponds, with their shallow water, are ideal for many species but may dry up for part of the year, so their inhabitants must be able to survive in a dormant state during such dry periods or move to other areas.

Lakes tend to have their own special animals and plants because they are in effect isolated places – islands of water in a sea of land. Most lake-living (lacustrine) species cannot travel along rivers, especially upstream when they would have to swim against the current and up waterfalls, so they are effectively cut off from the rest of the world. Lake inhabitants therefore tend to evolve their own characteristics and are often quite distinct from their counterparts elsewhere. Lake Baikal, for example, contains over 1,200 known species of animals and over 500 different plant species, and over 80 per cent of these are found nowhere else in the world.

Lakes and ponds, whether natural or man-made, contain relatively still water. In most rivers, the opposite is the case. Their animals and plants often have to adapt to strong currents; plants have to root themselves firmly to the river bed or rocks and many small river creatures have strong claws and suckers to do exactly the same. Swimmers have to expend a great deal of energy just to stay in one place. Even so, there are some real experts: salmon can travel great distances against the current, and torrent ducks happily swim with powerful strokes of their large, webbed feet in even the most frightening rapids.

In some rivers, even the experts cannot always develop methods of preventing themselves from being swept away. Particularly in mountain rivers, the currents are sometimes strong enough to move rocks and gravel. For this reason, many riverine species are confined to the lower reaches of the world's rivers where the water moves more slowly.

A typical river begins in a mountain as a small stream, though there are exceptions, such as the River Thames in England, which rises in the lowlands. The young river then passes through a range of different stages before it reaches the sea. These vary with the nature of the land it passes through and the activities of man along the way. Mountain rivers tend to have a great deal of oxygen in the water – the splashing and turbulence of the water flowing rapidly downhill over rocks helps to aerate it – which makes it suitable for animal life, but they also have dangerous torrents with waterfalls and rapids. Life is therefore harsh in the upper reaches. Where they broaden and meander through the lowlands, before meeting the sea in wide estuaries, the current is not so strong and life is generally easier.

Waterlilies *(Nymphaea sp.)*
above These specimens come from
Queensland, Australia.
Orange river, Namibia *left* A
rich habitat bordered by mountains.

Most freshwater habitats have been profoundly modified by man and even apparently isolated stretches of water have not escaped our influence. Their importance has been increasing ever since people first appeared on earth and began to fish and drink. Nowadays, in countries with a high standard of living such as Sweden, the USA and Britain, we use up to 500 litres (110 gallons) of water per day per person.   It is required for domestic consumption (cleaning, sewage removal, drinking and cooking), industrial purposes, irrigation for agriculture and many other processes. But in some parts of the world there is an increasing shortage of water, which is a serious problem because it cannot easily be traded from one country to another.

Not only are we using water itself but we are also modifying water habitats, for everything from hydro-electric power to fish farming. In addition the lower reaches of many major rivers are now confined within concrete embankments; they are also polluted by sewage, industrial waste and agricultural chemicals washed in from nearby land. Unfortunately many people look upon rivers as easy places to dump their domestic wastes, without looking at the tremendous ecological damage that results.

Perhaps ironically, in recent years there has also been increasing demand on lakes and rivers for recreational purposes. People like to visit them for sailing, fishing, bird watching, swimming, water skiing and many other activities. Yet even this kind of disturbance can cause serious problems unless properly controlled.

Freshwater invertebrates

Invertebrates such as insects and tiny crustaceans are the most prolific forms of animal life in ponds, lakes and rivers. They are far more numerous than the larger and better-known animals, in terms of both species and individuals, but even so, they are often overlooked. They are also able to take advantage of many different 'micro-habitats' offered by fresh water and live in the mud, between grains of sand, under stones, among plants, in the open water and in every other place you can imagine.

The daphnia is an excellent example of a small freshwater creature that is found almost everywhere but which is often overlooked. It belongs to a common and diverse group of small crustaceans known as the water fleas. Although barely more than one millimetre (one twenty-fifth of an inch) long itself, it feeds by filtering even smaller organisms out of the water.

The mosquito is also fairly small but, in contrast, is difficult to overlook. Not only is it incredibly common, it is also possibly the most dangerous animal in the world. Mosquito swarms are often so thick that people have suffocated when caught in them; worse still, these insects spread deadly diseases such as yellow fever and malaria. The male mosquitoes drink plant juices and are fairly harmless, but the females drink human and other mammal blood by piercing the skin with their modified mouth-parts (called 'stylets'); they are the ones which carry the diseases.

The mosquito's life cycle is exclusively linked to water. The female deposits her eggs either singly or in small clusters on the water or on an object nearby. The larvae – known as 'wrigglers' – swim by jerking their bodies but have to float to the surface at frequent intervals to push a slender breathing tube into the air. When fully grown they hang just below the surface and enter the pupal or resting stage; eventually the adult mosquito emerges from a split in the back of the pupal skin. It stands on the water for a while, until its wings are fully dried, and then flies away.

The pond skater spends its entire life 'walking' on the surface of the water. It has dense pads of waxy hairs on the ends of its legs which trap air and enable it to propel itself over the water, using special claws to get a grip, without actually breaking the surface film. The dents in the water surface where its feet rest can be seen on close examination. The middle pair of legs are moved in a rowing action, for propulsion; the hind pair steer; and the short front pair grab food. Pond skaters are predatory bugs that normally feed on dead and dying insects which fall onto the water. They stick their hollow proboscis inside the prey, like a hypodermic needle, and suck out the body fluids.

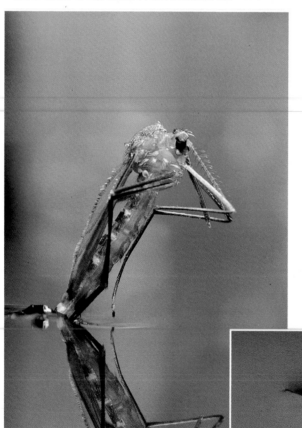

Daphnia magna *above*
Mosquito *(Culex pipiens) right*
Female emerging from pupa.
Pupa skater *(Gerris lacustris) far right*

Axolotl

Most newts and salamanders spend part of their time in fresh water and the remainder (usually outside the breeding season) on land. The axolotl (*Ambystoma mexicanum*), however, never leaves the water – because it spends its entire life as a tadpole.

Scientists have discovered that axolotls need the chemical iodine to complete their development into adults, but this substance is absent from their natural habitat. Therefore, in the wild they always stay as juveniles, although their sexual organs do mature eventually.

Restricted to Mexico's Lake Xochimilco and various canals in the vicinity, axolotls grow up to 25 centimetres (ten inches) long. Their name stems from an Aztec word meaning 'water monster'.

Axolotl (*Ambystoma mexicanum*) left

Alligator-snapping turtle

Weighing up to 100 kilograms (220 pounds) and as much as 75 centimtres (30 inches) in length, the alligator-snapping turtle (*Macroclemys temminckii*) is one of the world's largest freshwater turtles. Like many snapping turtles, it is an irritable animal and often shows great readiness to bite.

The turtle's jaws form a hooked beak rather like that of a bird of prey, but unlike its feathered counterpart it does not actively hunt for its food. Instead, this reptile prefers to lie on the bottom of a pond or lake with its mouth wide open, 'fishing' for passing prey. There is a bright red projection on the turtle's tongue, which stands out in the otherwise dark-coloured mouth and wriggles about like a live worm. As soon as an unsuspecting fish moves in to investigate, the turtle snaps its mouth closed and swallows it.

Salmon

Salmon are among the best-known and most prized fish in the world. Famous for their extensive and dangerous river migrations, they are marine fish which come to fresh water in the autumn to breed.

Every year enormous numbers of salmon swim great distances upstream against powerful currents, leaping as high as five metres (16 feet) over waterfalls, in order to return to exactly the same spot in the river where they hatched.

Having reached the upper stretches of the river, the female salmon lays her eggs in shallow depressions in the gravel beds and the male fertilizes them. The newly-hatched fish appear about three months later. They spend the first four weeks of their lives attached to the yolk-sacs of their eggs, until the food in the sac is used up, and then they feed on river plankton until they are about 18 months old. It is another three or four years before the silvery young salmon, now known as smolt, migrate to the sea. Some return to breed after only a year, others do not return for several years; but very few survive to make the journey twice. Most are too weak after breeding and die before they can complete the journey back to the sea.

MANATEE

DATA

SPECIES
West Indian manatee
(*Trichechus manatus*)

CLASSIFICATION
Sirenia (dugong and manatees)

DISTRIBUTION
South-eastern USA, Gulf of
Mexico, southern coastlines of
Caribbean Sea and Atlantic
Ocean to central Brazil, waters
around Bahamas and Greater
Antilles

HABITAT
Rivers and shallow, coastal
waters

SIZE
Maximum length 4.5 metres
(14½ft); weight up to 1.6 tonnes

FOOD
Aquatic plants

Manatees spend much of their time behaving exactly like underwater cows. Although quite unrelated, they graze on floating or submerged grasses and other plants and have become known in many parts of the world as 'sea cows' or 'fish cows'.

There are three species in the manatee group (*Trichechus* spp.): the West Indian manatee, the West African, and the Amazonian. They are similar in looks and behaviour and all three are among the most threatened aquatic animals in the world. Because they are docile and fairly tame, these air-breathing mammals have been hunted extensively for their meat and skins; they also suffer from the effects of pollution, flood-control dams and high-speed boats, which disturb and kill the sea cows with their propellers. Manatees produce only one calf every two years, so badly affected populations take a long time to recover even after they are fully protected. In some parts of the world there is now a more hopeful outlook for these harmless creatures. They are being introduced into irrigation canals and around the dams of hydroelectric power stations to clear weeds – so their future is perhaps a little more promising.

The calves are born underwater and are helped by their mothers to the surface, to take their first breaths. Usually newborn manatees can swim unaided within one hour. They begin to graze after a few months, though they still drink their mothers' milk for at least a year. Manatees are mostly solitary animals, though they sometimes form groups of a dozen or more in favourite areas. In clear water they explore by sight, but their eyesight is not very good and in murky water they often bump into things. Besides water vegetation they also occasionally eat invertebrates and small fish by mistake, taking them in on the plants without noticing.

Manatees are excellent swimmers. They can reach speeds of up to 25 kilometres (15 miles) per hour when pressed, and can dive underwater for as long as 15 minutes. They use their large, horizontal, paddle-shaped tails for swimming (by moving them up and down) and as rudders. The flippers are used in precise manoeuvring and to 'walk' along the bottom in shallow water.

There is a fourth species, very closely related to the manatees, known as the dugong (*Dugong dugon*). Distinguished by its different-shaped tail, the dugong is also threatened but can still sometimes be seen feeding on sea grasses in the coastal shallows of the south-west Pacific.

HIPPOPOTAMUS

The hippo spends most of its day resting or sleeping in the peaceful waters of lakes and rivers. It can remain underwater for as long as half an hour, but usually only sinks from sight for several minutes at a time. When it rises to take a breath its great head noisily breaks the surface, its eyes pop open and its ears unfold and wag. Then it rests with only the top of its head visible – just enough to see, hear and breathe.

Despite its enormous size (a bull hippo can weigh over 4 tonnes) and awkward looks, a hippo walks like a slow-motion ballet dancer along the bottom of calm rivers and lakes, following underwater paths. A hippo is so heavy that it does not float to the surface.

After sunset, all hippos leave the water to feed on land. Grass is their main food and they walk along narrow paths to their favourite grazing grounds, which may be as far as eight kilometres (five miles) away. In one night, a single hippo may eat as much as 70 kilograms (150 pounds) of vegetation.

If danger threatens these semi-aquatic mammals prefer to take refuge in the water – and they can run fast as a human over short distances if necessary. The adults are fairly safe from predators, though they have been extensively hunted by man for their highly prized flesh, the superior ivory of their teeth, and for 'sport'. They have also been persecuted in many areas because they enter cultivated fields and eat and trample crops. At one time, more than 60 hippos per kilometre (100 per mile) of river was not unusual, but they have now disappeared from many parts of their original range and their numbers have declined elsewhere.

At birth, which takes place in shallow water, the baby hippo weighs between 27 and 50 kilograms (60 and 110 pounds). It multiplies this weight by as much as five times during the first year. The youngster can swim before it can walk but often prefers to sit on its mother's back if crocodiles are nearby. In their first year young hippos are taken by leopards, lions, crocodiles, hyenas and wild dogs.

At one time it was thought that hippos sweated blood. This fallacy arose because there is a red, oily liquid which oozes from pores in their skin. It is designed to keep the skin moist and to kill germs and heal wounds but it can look very similar to blood. In fact fights and injuries among hippos are common. Adult males compete with one another for the control of their herd and for territories in vicious fights that often last several hours. They use their canine teeth – which can grow to 60 centimetres (two feet) in length – and have been known to kill one another in really desperate battles.

D A T A
SPECIES
Hippopotamus
(*Hippopotamus amphibius*)
CLASSIFICATION
Artiodactyla (pigs, cattle, deer and others)
DISTRIBUTION
Africa south of Sahara, though rare or absent in many original haunts; highest densities in East and Central Africa
HABITAT
Lakes and rivers with deep water and adjacent grassland
SIZE
Up to 4.5 tonnes; 1.5 metres (nearly 5ft) at the shoulder
FOOD
Grass

FISHING BAT

D A T A

SPECIES
Fishing bat
(*Noctilio leporinus*)

CLASSIFICATION
Chiroptera (bats)

DISTRIBUTION
Southern Mexico to northern
Argentina; Jamaica, Cuba,
Hispaniola, Lesser Antilles

HABITAT
Ponds, slow-moving rivers,
sheltered coastal lagoons

SIZE
Head and body length up to
13cm (5in)

FOOD
Small, surface-swimming fish;
some insects, including scarab
beetles and stinkbugs

There are no fewer than 951 different bat species around the world. Found everywhere except in the Arctic and Antarctic and on the highest mountains, they range in size from the tiny kitti's hog-nosed, or bumblebee, bat which has a wingspan of only 15 centimetres (six inches), to the enormous flying foxes with wingspans of two metres (over six feet) or more.

As a group, bats have exploited virtually every kind of land habitat and consume an enormous range of different foods. Several species have even learned to catch fish and two in particular are so well adapted to this way of hunting that they have become known as the fishing bats.

Always found near rivers or other sources of water, these two species (*Noctilio leporinus* and *N. albiventris*) live in Mexico, Central America and the northern half of South America. They can also be found on Cuba, Hispaniola, Jamaica and the Lesser Antilles. The smaller *N. albiventris* also takes insects, but its larger relative eats fish almost exclusively. It generally hunts at dusk and dawn, over large rivers and lakes and occasionally out over the sea. Like most other bats, fishing bats have a marvellous ability to avoid obstacles and catch food in total darkness. By sending out series of short, high-pitched beeping sounds (usually beyond the range of human hearing), and then listening for the echoes that bounce back when the beeps hit an object, they can tell where the object is and whether or not it is moving. As they hunt they simply skim across the surface of the water in a zigzag flight pattern and wait until the beeps identify a fish breaking the surface nearby.

The most characteristic feature of a fishing bat is its huge feet and the sharp claws on each toe. The bat hooks the fish, which may be as long as eight centimetres (three inches), with these claws and quickly lifts it to its mouth. The fish is sometimes eaten in flight but more often carried to a roost in a hollow tree, rock cleft, cave or old building. The technique is highly successful, though the bats occasionally misjudge a swoop and fall into the water, extracting themselves by using their wings as oars to leap out again. It has been estimated that on an average night each individual catches between 30 and 40 fish! When its normal prey is scarce, the fishing bat uses the same system to catch a range of insects.

DUCK-BILLED PLATYPUS

The duck-billed platypus is such a strange-looking animal that it took a long time to convince people that it really exists. Originally, when the first skin arrived in Britain from Australia in 1798 scientists thought it was a fake; they even accused a taxidermist of stitching together several parts of different animals. There were webbed feet and a bill like a duck, but a fur covered body and a flat tail like a beaver. Later the scientists were even more amazed to learn that the duck-billed platypus is one of only two kinds of mammal that lay eggs (the other is the echidna or spiny anteater).

The duck-billed platypus lives in the fresh-water streams, lakes and lagoons of eastern Australia. Its duck-like snout (which is not hard but soft and leathery) has evolved for scooping up aquatic worms, shellfish, insects, insect larvae and other small animals from the bottom. These bits of food are stored in the cheek pouches until the platypus rises to the surface to chew and swallow them.

Platypuses live in burrows that they dig in the bank, with the entrance just above the water surface. The females also dig special nesting burrows in which to lay their eggs. Typically these consist of a winding tunnel, usually about eight metres (26 feet) long – though tunnels up to 30 metres (100 feet) in length have been recorded – and a small chamber, lined with grass and leaves, at the end. The female seals herself inside the chamber with a plug of earth and lays two or three pale, sticky, leathery eggs, each only one and a half centimetres (half an inch) long. She keeps them warm between her body and tail until they hatch about ten days later. The newborn platypuses are hairless and tiny, measuring about two centimetres (less than one inch) long, and they emerge from the burrow only after three or four months of suckling, in the late Australian summer. Like their parents, although they shuffle awkwardly on land the babies are very graceful in water. Their eyes and ears close underwater so they are blind and deaf, but the bill and head hairs have an excellent sense of touch which they rely on enormously.

Young platypuses all have sharp spurs on the heels of their hind feet, though in the females these soon disappear. The males can inject a poison through their spurs and use them to jab enemies in defence. The poison causes agonizing pain in people and can kill an animal the size of a dog.

Platypus pelts were once highly sought-after for the fur trade and at one time the animals were on the brink of extinction in many areas. Protective measures have been enormously successful, however, and they have made a comeback in most parts of their range.

DATA	
SPECIES	Duck-billed platypus (*Ornithorhynchus anatinus*)
CLASSIFICATION	Monotremata (egg-laying mammals)
DISTRIBUTION	Eastern mainland Australia (Queensland, New South Wales, Victoria) and Tasmania
HABITAT	Freshwater streams, lakes and lagoons
SIZE	Up to 60cm (2ft), including tail of 15cm (6in)
FOOD	Worms, shellfish, insects, insect larvae and other small aquatic animals

WHITE PELICAN

DATA

SPECIES
White pelican
(*Pelecanus onocrotalus*)

CLASSIFICATION
Pelecaniformes (pelicans,
cormorants, gannets and
others)

DISTRIBUTION
East and southern Africa, south-
eastern Europe, south-western
USSR, parts of western Asia

HABITAT
Swamps, open water and
coastlines

SIZE
Head and body length 1.6 metres
(5ft); wingspan 2.5 metres (8ft)

FOOD
Fish

Pelicans are famous for their long bills and enormous throat pouches. Contrary to popular opinion, the pouch is not used to store fish but is designed as a scoop to catch them. Once trapped in this manner, the excess water is strained off and the fish are quickly swallowed before the bird tries to catch more. Groups of pelicans can often be seen working in teams, sitting on the water and herding together shoals of fish. In these 'fishing schools' all the birds submerge their heads and necks and scoop up the fish at exactly the same time.

The exception to this method of fishing is the brown pelican (*Pelecanus occidentalis*), which dives into the water from high above, like a gannet. As soon as a suitable fish is spotted, this pelican folds its wings to increase its falling speed and hurtles down into the sea to dive on the unsuspecting prey.

There are seven species of pelicans around the world, divided into two main groups. The first consists of predominantly white birds which nest on the ground; the other comprises mainly grey or brown birds which nest in trees. Both groups prefer undisturbed areas in which to nest, preferably with plenty of fish nearby. However, they are all good fliers and very skilful at using thermal updrafts to conserve energy, so they often commute hundreds of kilometres every day to lakes which offer good fishing.

Although the female pelican lays several eggs, with two- or three-day intervals between each, the pair usually manage to raise only one bird. This is because the first young pelican to hatch has the advantages of age and size when feeding, so it takes most of the food; often the parents cannot supply enough to go round and the younger, smaller nestlings starve to death.

White pelicans are common birds, sometimes living in colonies of 40,000 or more. But Dalmatian pelicans are down to between 500 and 1,400 breeding pairs, in only 19 colonies between eastern Europe and China. They have suffered extensively from human disturbance, predation and flooding problems. The grey pelican is also in trouble, with only 1,300 breeding pairs in a total of 27 colonies in India and Sri Lanka – the result of pesticide poisoning, habitat destruction and, once again, human disturbance.

DARTER

Many people mistake darters for snakes in the water, because they often swim partly submerged with only their heads and long, thin necks above the surface. Darters are often nicknamed 'snake-birds' because of this resemblance, though they are actually related to cormorants and pelicans.

There are four species of darter, found in the warm waters of America, Africa, Asia and Australia. All are very similar in appearance and way of life, preferring quiet waters with plenty of surrounding trees and islets on which to build their nests.

When nesting, darters are more territorial than either cormorants or pelicans. The male selects a nestsite among the boughs of a tree and claims a territory around it, which often includes the whole tree. He decorates the site with a few fresh, green, leafy twigs, and displays with his wings and the twigs to attract a mate and to ward off other male darters. His new partner then builds the nest, mostly with material which he collects, in the fork of a tree branch over the water. Both parents guard the eggs and young, taking turns and usually doing one or two shifts each per day. While one parent remains at the nest site, the other catches fish to feed the young. In fact the young chicks are fed with a soup of partly digested fish.

Most water birds maintain their buoyancy and keep afloat using air trapped within their waterproof plumage. Darters can change their buoyancy at will in order to swim either at or beneath the surface. They do this by flattening their feathers against their body to expel the trapped air. In addition the darter can empty its 'air sacs', the special balloon-like extensions to the lungs that all birds possess. The function of the air sacs is not fully understood but is believed to be concerned with flight; the darter's flying abilities do not seem to suffer, however, and with the air in its plumage and its air sacs gone it can sink like a stone.

Darters are fast swimmers underwater. They propel themselves with their webbed feet, aided to some extent by their wings. While hunting the neck is held in a coiled S-shape, ready to dart forward and snap up insects on the surface or stab at fish. When fishing the beak is held slightly open and used like a spear, to stab the animal in the side. It is then shaken loose, flipped into the air and swallowed head first. Many an unsuspecting predator and human molester has also suffered from these lightning stabs, which are often used in self-defence.

The mechanism involved in the strike consists of two special joints between the vertebral bones in the darter's neck. Before the strike the neck is held in the 'S' shape by long, sinewy tendons running over a pulley system formed by these bones. Powerful muscles then straighten the neck and the head snaps forward.

The characteristic S-shaped 'kink' in the neck is also conspicuous when the darters are resting, making them relatively easy birds to identify. The combination of the long neck and also the unusually long tail, for waterbirds, makes the darter look distinctive in flight as well – almost like a flying cross.

DATA
SPECIES
Darter (*Anhinga rufa*)
CLASSIFICATION
Pelecaniformes (pelicans, cormorants, gannets and others)
DISTRIBUTION
Africa south of the Sahara, southern Asia, New Guinea
HABITAT
Lakes, lagoons and rivers
SIZE
Length 80 to 85cm (31 to 33in)
FOOD
Fish; occasionally insects and other water creatures

ISLANDS

Jamaica, Hawaii, Ascension, Iceland, Madagascar, Tasmania, Sri Lanka, Mauritius, South Georgia and Majorca: some are just tiny dots in the ocean, others are several thousand kilometres long, but all are islands. Many are a stone's throw from the nearest mainland, while some are hundreds or thousands of kilometres from the nearest land, in the middle of vast seas.

Most of the islands which are near the mainland were originally joined as part of a peninsula, but have been cut off from the mainland by the wind and waves slicing a channel between them. Remote islands are usually volcanic in origin, being the tops of underwater mountains jutting above the surface of the sea.

As well as geographical islands there are also 'ecological islands' which show very similar characteristics. These are not cut off by stretches of ocean but they are isolated just as effectively from similar habitats by other, ecological, barriers. Among these ecological islands are fragmented pieces of rainforest. Far from being in large unbroken tracts, the world's tropical rainforests have been destroyed to such an extent that the small areas that are left have effectively become islands separated by enormous 'seas' of agricultural land or desert.

Many lakes are also cut off, this time from other stretches of water, and so are 'watery islands' in a sea of land. Lake Baikal, in central Asia, is the oldest, deepest and most remote large lake on earth. It has been isolated from all other lakes for more than 30 million years.

Most islands inherit many of the animals and plants which were already present before they were cut off. But species living on volcanic islands, which rose barren and lifeless from the sea, have arrived of their own accord. It is incredible how many plants and animals have been able to travel such enormous distances. Indeed, in time, living things seem to be able to reach even the most remote islands in the world.

Birds are such accomplished travellers that they tend to be the first to arrive. Albatrosses, for example, come across new islands in the course of their ocean patrols and use them initially for breeding, or simply for resting. Smaller birds, prone to being blown off course by strong

Heron Island *above*
Situated on the Great Barrier Reef, this island's scenery is typical. The trees in the foreground are Pandanus palms.

Coco de Mer *right*
These specimens come from the Seychelles.

winds while on migration, arrive mostly by accident.

Many plants are also good travellers. Coconuts can remain alive floating at sea for as long as four months and are often carried hundreds of kilometres by currents or winds before being cast up on a new beach. Other plant seeds, and many insects, are literally blown over the sea by the wind or are carried by birds, either stuck to their beaks, feet or feathers or within their stomachs, in which case they get squirted ashore in droppings.

While a few larger animals such as turtles and seals are able to swim, smaller creatures have to rely on floating logs or mats of vegetation for their ocean voyages. These 'rafts' are sometimes swept down to sea by tropical rivers in flood and themselves become islands, floating along with snails, millipedes, spiders and many other animals and plants as passengers. Reptiles are particularly hardy sailors, capable of surviving long voyages on such rafts. Many smaller species, intent on munching their way through the leaves and other food that is travelling with them, may be unaware of their changed circumstances.

Life ashore

Once ashore, those that have coped with the hazards of the sea inevitably need to adapt to their new surroundings if they are to survive. Each and every living creature is slightly different from all the others and those with differences that are useful in the new environment are the ones most likely to be able to survive and breed. If the useful differences are passed on to their offspring, each time being slightly exaggerated, then many generations later the variations become so great that a new species has been formed. This is the basis of evolution.

When isolated islands are first inhabited, this species formation occurs very rapidly because the survivors have to adapt quickly to the new conditions. For this reason, and because they are evolving in isolation from species anywhere else in the world, island animals and plants tend to be quite distinct, often unique.

There are two trends which distinguish many island species from those found on the mainland. Firstly, since there is less competition on islands for food they tend to grow larger. Island reptiles are often huge, such as the famous giant tortoises of the Galapagos Islands and the Seychelles. The Komodo dragon of Indonesia is another example: at three metres (10 feet) long it is considerably bigger than any other lizard in the world.

The second trend is that, since there are generally fewer predators on islands than elsewhere, there is no need for many species to be able to hide or flee. As a result many island birds, for instance, have become completely flightless. The elephant birds of Madagascar, which are now extinct, illustrate both these trends. Not only were they unable to fly but they were also the largest birds that have ever lived.

However, it is exactly this kind of extreme specialization which makes island species especially vulnerable to outside influences. The fact that the elephant bird was both large and unable to fly ultimately led to its extinction, because it was unable to cope with the invasion of people when they arrived on Madagascar some 1,500 years ago. The dodo suffered a similar fate. Visitors to its island home of Mauritius, a few centuries ago, treated this flightless and trusting bird as little more than a convenient source of food. The island's dodos had no time to adapt to the presence of·man, a new and particularly voracious predator, and the last one was killed in 1681 – less than 200 years after the species had first been discovered. Today only 20 per cent of all the world's birds are island species, but no less than 90 per cent of all those driven to extinction in recent times have been island dwellers.

People, of course, are the most destructive form of outside influence which any island has to face. Not only do humans find the large, curious and often fearless wildlife easy and welcome prey, but they also bring with them their domesticated animals such as rats, goats, pigs, cats and dogs. For many native species these new introductions are the first predators and competitors they have had to face. Not surprisingly, the results are usually disastrous.

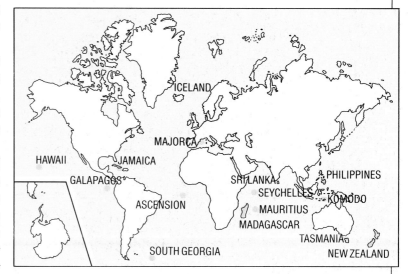

The world's islands above
This map pinpoints some of the islands of the world that are most important from the point of view of wildlife, from Iceland in the North to South Georgia.

Madagascar

Madagascar is the fourth largest island in the world and has been separated from the south-east coast of Africa for about 60 million years. It has one of the most unusual assemblies of wildlife anywhere on earth. No less than nine out of every ten of its inhabitants are found only on this island. There are hedgehog-like tenrecs; 150 species of spectacularly coloured frogs, all except two of which are found nowhere else in the world; a cockroach that hisses when angry and whistles when in love; and a spectacular collection of flowering shrubs, lianas, orchids and other plants. More than 10,000 different plant species have been identified so far, and botanists estimate that there could be 3,000 or more yet to be discovered.

Lemurs are perhaps the most exciting and best known of all Madagascar's wildlife. There are about 20 species, some as small as mice and others as large as orang-utans. Among them are sifakas (*Propithecus verreauxi*) which sit on the treetops, arms outstretched to worship the morning sun; ring-tailed lemurs (*Lemur catta*) in places so tame and sociable that they seem to be more common than they really are; and indris (*Indri indri*) with their melodious and beautiful calls which echo for kilometres throughout the forests.

One very strange and rare species is the aye-aye (*Daubentonia madagascariensis*). Resembling the famous film character 'ET', though in residence millions of years before its science-fiction look-alike appeared, the aye-aye is a peculiar collage of animal parts. The size of a house cat, it has ears like a bat, beaver-like incisors, a tail like an overgrown ostrich feather, owlish eyes and a skeletal middle finger like a dead twig. Once thought to be extinct, it now just survives on an uninhabited island reserve off Madagascar's north-east coast.

Unfortunately, there are many other animals and plants threatened with extinction on Madagascar. Ever since people first arrived they have been chopping and burning down the rainforests to provide more room for crops and cattle. Now only a small area of forest is left. As a result, many species have become extinct and many more are in serious danger of disappearing in the near future.

Aye Aye

ENDANGERED

(*Daubentonia madagascariensis*)
top right

Ring-tailed lemur (*Lemur catta*)
below

Sifaka (*Propithecus verreauxi*)
above

Sealions of Galapagos

The Galapagos sealion (*Zalophus californianus wollebaeki*) is a common animal on and around the Galapagos archipelago. Preferring smooth, sandy beaches, it occurs on all islands of the group and breeds on most. It is actually a sub-species of the better-known Californian sealion, which is a popular animal in zoos and circuses and as its name suggests occurs mainly in Californian waters. There is also a third sub-species, the Japanese sealion, which is found in Japanese waters – although there have been no verified records since the Second World War.

Although their natural history is not as well known as it is for their relatives on the mainland, Galapagos sealions are thought to be opportunistic feeders. Their stout whiskers are probably used to detect disturbances in the water caused by such prey as fish, squid and octopus.

Galapagos sealions are very playful animals. Under the water they have been seen chasing and catching their own exhaled bubbles of air. This is interesting because unlike true seals, which breathe out and empty their lungs before diving, sealions dive with some air in their lungs. The air is used by the males to call underwater when they are patrolling their territories.

Galapago sealion (*Zalophus wollebaeki*) left

Island crabs

There are about 6,000 different species of crabs around the world. They range in size from the enormous Japanese spider crab (*Macrocheira kaempferi*), which has a pincer-span of over two metres (seven feet), to the tiny parasitic pea crabs (Pinnotheridae) with a shell diameter of only six millimetres (one quarter of an inch). They all belong to a diverse group of animals known as the Crustacea, which also includes barnacles, shrimps, prawns, lobsters and woodlice.

Nearly all crabs are sea-dwellers, able to live out of water for only short periods. But there are some air-breathing species such as the red rock or sallylightfoot crab (*Grapsus grapsus*) which is able to remain out of water for several hours. The sallylightfoot is only found on the rocky shores of the Galapagos Islands in the Pacific, though crabs are generally uncommon or absent from remote islands in other parts of the world. Those crabs which are found on Pacific Islands, and in places like Trinidad, off Brazil, are of great interest to scientists studying evolution and adaptation. This is because, although mainland tropical shores teem with dozens of different crab species, each remote island has been colonized separately and so has its own species, suited to its surroundings and found nowhere else in the world.

Sallylightfoot crabs (*Grapsus grapsus*) below

MARINE IGUANA

DATA

ENDANGERED

SPECIES
Marine iguana
(*Amblyrhynchus cristatus*)

CLASSIFICATION
Squamata (snakes and lizards)

DISTRIBUTION
Galapagos Islands

HABITAT
Rocky coasts of several different
islands

SIZE
Up to 1.75 metres (5½ft) long

FOOD
Algae, particularly seaweeds

There are about 3,000 species of lizard in the world, but only one is truly at home in the sea. The marine iguana of the Galapagos Islands often swims from one island or islet to another and is quite happy underwater, swimming quickly and easily with its legs motionless and a simple snake-like movement of its body and tail. It excretes the excess salt it takes in through special glands in its nose.

Despite their large size marine iguanas are harmless, peaceable animals and show no fear when approached. They emerge at first light from crevices in the black lava rock along the coast. Their day is divided between sunbathing, since they are cold-blooded and must do this to soak up necessary body heat, and lying or feeding in the surf. Like many other iguanas they are herbivorous, but have developed a unique way of consuming food. They eat seaweed, which they find both above and below the water surface, taking bites alternately with the right and left sides of their mouths – rather like a dog chewing on a bone.

The greatest concentrations of marine iguanas occur on protected lava reefs, spits and shorelines, where there is plenty of seaweed for feeding and also adequate soil just inland for nesting. During the mating season the males hold court on projecting rocks, each one often gathering several females around him. Opponents test their strength by butting their heads together and trying to push each other away from the field of battle. For the rest of the year they lie happily side by side and on top of one another, often in groups of several hundred at their favourite locations, soaking up the blazing sun.

Although there are quite healthy populations of the known sub-species particularly on the islands of Albermarle and Narborough, the marine iguana numbers are decreasing in some areas and especially near human habitation. Early mariners used to eat them and many were killed near human settlements, but today the greatest threats come from feral dogs, cats and other introduced animals, which eat the young and sometimes even the adults.

KOMODO DRAGON

In the last century there were rumours of giant 'terrestrial crocodiles' living on the island of Komodo, in Indonesia. There were tales of these beasts, seven metres (23 feet) or more long, waiting in ambush for unwary people walking along game trails. Then, in 1912, the Komodo dragon was discovered by Western scientists. It turned out not to be a crocodile at all, and seven metres was found to be an enormous exaggeration of its length, but the discovery created a sensation.

The Komodo dragon is the world's largest lizard. It lives only on Komodo, a mere 30 kilometres (19 miles) long, and a few other small Indonesian islands: Padar, Rintja and the western end of Flores. The adult males grow up to three metres (10 feet) long and possess powerful teeth one centimetre (two-fifths of an inch) long with jagged rear edges. Nevertheless stories of these 'dragons' hunting people are almost certainly untrue, though they may bite viciously when frightened or cornered. In captivity, they can become quite tame.

The Komodo dragon belongs to a group of lizards known as monitors, which includes about 30 species ranging in size from 20 centimetres (eight inches) upwards. All of them are active by day and usually hunt by sight. When pursuing their prey the monitors lift their tails off the ground and run on their powerful legs, which makes the larger ones in particular look unnervingly like the legendary dragons of olden times.

Like all monitors, the Komodo dragon is a meat-eater. It often takes carrion and is cannibalistic at times (the adults regularly eat smaller individuals). But most of all, it is a hunter. The young animals prey on insects and small lizards, often climbing into trees to get them, while half-grown individuals hunt mostly on the ground for rats, mice and birds. The adults live on wild pigs, deer and goats, grabbing them in their vice-like jaws and shaking them violently until they die. Komodo dragons and other monitors are the only lizards that can increase the size of their mouths by dislocating the lower jaw – the same adaptation as in snakes – which enables them to carry out the amazing feat of swallowing large prey whole.

In the past Komodo dragons have been commercially exploited for their skin. Today the greatest threats are direct persecution (to prevent them preying on domesticated pigs and goats), ever-decreasing numbers of wild pigs and deer because of over-hunting by man, and habitat destruction. There are now barely 6,000 individuals left and the Komodo dragon is threatened with extinction in some areas.

DATA

ENDANGERED

SPECIES
Komodo dragon
(*Varanus komodensis*)

CLASSIFICATION
Squamata (snakes and lizards)

DISTRIBUTION
Komodo and a few other small neighbouring islands in Indonesia

HABITAT
Island savanna (grasslands and bushy areas with scattered trees)

SIZE
Up to 3 metres (10ft) in length; weight 135kg (300lb) or more

FOOD
Adults feed on wild pigs, deer and domesticated goats; young on smaller animals such as insects, rats and mice

JAMAICA BOA

DATA
ENDANGERED
SPECIES Jamaica boa (*Epicrates subflavus*)
CLASSIFICATION Squamata (snakes and lizards)
DISTRIBUTION Jamaica, and possibly Goat Island, in Caribbean
HABITAT Scrubland with rocky areas
SIZE Approximately 2 metres (7ft) long
FOOD Probably rodents, birds and lizards in the wild

Many adventure stories set in the tropics contain descriptions of enormous snakes lying in wait in the trees and then wrapping themselves around people with lighting speed. But only an exceptionally big snake, perhaps 10 metres (33 feet) long and with a very large circumference and weight, would be able to subdue a person. Very few species, all belonging to the boa family, can attain this necessary size: the anaconda (*Eunectes murinus*), the boa constrictor (*Boa constrictor*), and one or two others. Even with these large species there have been very few cases of people being attacked and eaten.

The Jamaica boa, or yellow snake as it is often called, is closely related to both the anaconda and boa constrictor, but it is comparatively small and rarely seen. Very little is known about its life in the wild. It is thought to feed on rodents, birds and lizards, killing its victims by coiling round them and squeezing to produce suffocation, then swallowing them whole. Like all boas, this species has no poison fangs and can only kill by constriction.

As its name suggests the Jamaica boa lives on the island of Jamaica in the Caribbean, but it is also thought to exist on nearby Goat Island. Mostly active at night, it spends its time in scrubland, rocky areas and cultivated land. It breeds every other year and 15 to 20 (though sometimes as many as 40) young are born.

Jamaica boa numbers have fallen drastically in recent years. The main threat is from introduced mongooses but the boas are also preyed on by feral cats. As with most snakes they are killed on sight by local people, and due to the species' rarity they are also subject to collection for the pet trade – though this is now illegal.

The Jamaica boa is not alone among island boas in being threatened with extinction. The Puerto Rican boa is also suffering from predation by mongooses and at one time was killed in large numbers for its valuable oil; the Round Island boa, if not already extinct, is restricted to Round Island, a 151-hectare (372-acre) islet to the north-east of Mauritius; and the Cuban boa is declining fast as a result of forest destruction.

GIANT TORTOISE

Giant tortoises were once so common in the Galapagos Islands that people claimed they could walk long distances over their shells without touching the ground. They certainly used to exist in massive numbers but these have been reduced considerably in recent centuries. So many were slaughtered for their meat, by pirates and whalers from passing ships, that some estimates indicate no fewer than ten million have been killed in the space of a few hundred years. More recently they have suffered from competition for food with goats, donkeys and cattle which have been introduced by settlers.

There are about 14 different sub-species of giant tortoise in the Galapagos Islands (a separate species also lives in the Seychelles) and some of these are already extinct. One particular variety, *Geochelone elephantopus abingdoni*, has only one known living representative. Affectionately known as 'Lonesome George', he now lives at the Charles Darwin Research Station, where important captive breeding and rearing projects are being carried out with his endangered close relatives in the hope that, eventually, they can be returned to their island homes.

There are surprising differences between each kind of Galapagos tortoise and this is one of the factors which led Charles Darwin to his famous theory of natural selection. No two kinds can be found on the same island, which means that they have all taken on different characteristics as a result of evolving independently. The only exception is the island of Isabela, where there are five sub-species, but these are separated by impassable volcanic ridges and peaks which are as effective barriers as the sea.

The subspecies differ mainly in the shape of the shell, or carapace, although the length and thickness of the legs and heads also varies. In general, those subspecies consuming tall-growing plants have a 'saddleback' shell and long neck, presumably so that the animals can stretch up to reach their food. A giant tortoise will eat almost any type of plant food and readily adapts to raiding kitchen gardens.

These enormous reptiles breed at any time of the year. The males, who are larger than the females, overpower their partners and laboriously haul themselves on top of the females' shells to mate – an amusing slow-motion spectacle. Then each female moves to a valley area and selects a patch of bare earth which receives plenty of sun (for incubating the eggs), urinating on the hard earth to soften it before digging a pit about 30 centimetres (one foot) deep. Around 10 to 15 eggs are layed and covered with soft soil. As the young hatch they dig themselves out of the earth and quickly hurry into cover in order to hide from predators.

Giant tortoises grow up to one and a half metres (five feet) long and can weigh more than 200 kilograms (440 pounds). They live longer than any other animal – possibly to as much as 200 years – which means that some individuals alive now were around in Darwin's day. They usually live in warm, dry lowlands but regularly trek along well-worn paths far away into the volcanic highlands, in order to wallow and drink in the muddy pools. These paths have been used for centuries and, if conservation efforts are successful, they will be used for centuries to come.

DATA

ENDANGERED

SPECIES
Giant tortoise
(*Geochelone elephantopus vandenburghi*)

CLASSIFICATION
Chelonia (turtles and tortoises)

DISTRIBUTION
Caldera and Isabela islands in the Galapagos archipelago

HABITAT
Warm, dry lowlands; spends some time partially submerged in pools in highlands

SIZE
Up to 1.5 metres (5ft) long; some individuals weigh over 200kg (440lb)

FOOD
Almost any kind of green vegetation

DARWIN'S FINCHES

DATA

SPECIES
Darwin's woodpecker finch
(*Camarhynchus pallidus*)

CLASSIFICATION
Passeriformes (perching birds)

DISTRIBUTION
Galapagos Islands

HABITAT
Commonest in higher zones; also
occurs in arid areas

SIZE
15cm (6in)

FOOD
Large insects underneath bark
and in soft, decaying wood

KAKAPO

ENDANGERED

The flightless kakapo (*Strigops habroptilus*) is unique among parrots. Only capable of wing-assisted leaps or parachuting, it is a nocturnal bird and likes to spend the daytime hiding in a secluded, well-shaded corner. It is also the world's heaviest living parrot, weighing up to three and a half kilograms (seven and a half pounds), and it has an inflatable air sac which it uses to make an extraordinary booming call similar to the sound made by blowing across the top of an empty bottle. There may be fewer than 50 kakapos left in the wild, surviving in only two remote areas of New Zealand. They were popular to eat at one time and their feathers were highly prized for Maori cloaks and to fill mattresses. To add to their problems many have been eaten by the wild cats, dogs and rats.

There are relatively few land birds in the Galapagos Islands, but the majority present are both unique and unusual. One particular group, known collectively as Darwin's finches, have excited tremendous attention over the years even though they are rather drab and inconspicuous birds.

Scientists believe that when the Galapagos archipelago was first being colonized, only one kind of finch managed to reach the islands from the mainland. The small group of birds distributed themselves among the islands, and eventually different populations became adapted to their different environments. Today they have evolved into 13 different species, all sparrow-sized or slightly smaller. Many are superficially similar, if not identical, and therefore extremely difficult to tell apart. They all have poor powers of flight, presumably due to the relative freedom from predators and the high risk of being blown out to sea and lost if they fly too high or venture too far afield. Even their breeding habits and songs are almost identical. Indeed, the song often varies more within any one population than it does between different species.

The only way of distinguishing the finch species, and even then it is difficult, is by their beaks. Each species' beak has become adapted to cope efficiently with one particular type of food. For example, the thin, pointed beak of the warbler finch (*Certhidea olivacea*) is for picking insects off leaves – and incidentally the bird itself looks, behaves and sings like a real warbler. The large ground finch (*Geospiza magnirostris*) has a big, tough bill for cracking open hard seeds. The woodpecker finch (*Camarhynchus pallidus*), which is perhaps the most remarkable of all, uses its bill for probing into dead wood or cracks in the bark. It sometimes even uses a twig or cactus spine for poking out otherwise unobtainable prey, in much the same way that a real woodpecker uses its long, probing tongue.

For many years Darwin's finches were thought to be restricted to the Galapagos Islands, but in 1891 another species was found to inhabit Cocos Island, about 900 kilometres (560 miles) to the north-east. Overall the finches are very common birds throughout the entire Galapagos archipelago. It is because of their very different beaks that they have been able to minimize competition by specializing on various parts of the limited foods available to them.

Woodpecker finch *below*
This finch, photographed on Santa Cruz Island, Galapagos, is using a cactus spine as a tool.

MULLER'S PARROT

FAIRY TERN

The true fairy tern (*Sterna nereis*) has a black cap of feathers on its head and nests on the coasts of Australia, New Caledonia and northern New Zealand. However, there is another species, the white tern (*Gygis alba*), which is often given the same name because it is a beautiful snow-white in colour except for its dark bill and eye rings. The better-known white tern breeds in the Seychelles and many other tropical islands. It lays its single egg precariously balanced on a tree branch, where it is in constant danger of being knocked to the ground. The newly hatched chicks also have balancing problems and have enormous, rather comical, hooked feet for hanging on to their perches. They patiently cling to the same branch on which they hatched, waiting for their parents to bring them fish, for several weeks before attempting to fly.

Parrots are renowned for their colourful feathers. The Muller's parrot is no exception, even though it is not one of the brightest of the 328 species in the family. Found in the Philippines and on Celebes and nearby islands in Indonesia, it inhabits the edges of forests and clumps of trees in cultivated land. It is very rarely seen in the dark forest interiors.

Muller's parrots are active at night and more often heard than seen. Their screeching, a harsh 'ki-ek . . . ki-ek . . . ki-ek', is unmistakable as they fly overhead. They feed mostly on fruit, seeds, nuts and berries (taken in the treetops), but also make themselves unpopular by attacking ripening corn. This is possibly their favourite food and they reportedly cause considerable damage to corn crops.

Like all parrots, the Muller's most obvious characteristic is its bill. It is an extremely adaptable structure which can be used to perform delicate tasks such as preening but at the same time is powerful enough to crush the hardest nuts and seeds. The hooked upper half of the bill is attached to the skull by a kind of hinge which helps the bird to use it as an extra 'foot', like a grappling hook for clambering about in the treetops. Likewise its feet, which are specially adapted for grasping with two toes in front and two behind, give an extremely powerful grasp or they can be used for delicately holding and manipulating objects close to the bill.

Although as a group parrots are comparatively successful animals, many species have become extinct within the last few centuries and many more are seriously endangered. Island species are especially vulnerable, since most have small populations and slow breeding rates and are less able to cope with habitat destruction and introduced competitors, predators and diseases. The most serious threats to parrots everywhere include collection for sale to zoos and as pets, persecution as agricultural pests, and destruction of the tropical rainforests which are the preferred habitats for most species. The Muller's parrot has so far survived in fairly good numbers, but endangered island species like the Puerto Rican parrot, the St. Lucia parrot and the St. Vincent parrot have not been so lucky.

DATA	
SPECIES	Muller's parrot (*Tanygnathus sumatranus*)
CLASSIFICATION	Psittaciformes (parrots, parakeets and cockatoos)
DISTRIBUTION	Philippines and Indonesia
HABITAT	Edges of island rainforests and clumps of trees in cultivated land
SIZE	32cm (13in)
FOOD	Fruit, seeds, nuts and berries

COASTLANDS

Allied to both land and sea, yet belonging to neither, the world's coastlines support a great abundance of life. Sandy beaches, rocky shores, sea cliffs, shingle spits, sand dunes and rock pools are all coastland habitats. They are the ribbons between land and sea, that surround every island and every continent across the world.

The type of coast found in a particular area depends on many factors. Hard rocks tend to form a cliffed and rocky shore, while soft ones erode to give sloping sandy or muddy shores. But the effectiveness of the waves in wearing down the rock also has a strong influence, and in addition the local wildlife depends on the temperature of the water and surrounding air, on the strength of the currents and on many other factors – including the activities of man.

All coastlands are continuously evolving by the processes of erosion and deposition. Some, such as sand dunes, may form within decades, while others, such as granite headlands, show little change over many centuries. Cliffs formed from soft rocks are especially liable to serious and rapid erosion and the material abraded from them is often transported along the coasts to form sandy beaches elsewhere.

One of the main characteristics of almost all coastlines is the presence of tides. Tides are the alternate rising and falling of the world's oceans and seas, governed by the gravitational pulls of the Sun and the Moon. They vary throughout the world, but in most places there are two high tides and two low ones roughly every 24 hours. Also, twice a month there are 'spring tides' when the Sun, Moon and Earth are in an approximate straight line and the water surges very high up the beach and then retreats a long way down as it goes out. Alternating with the spring tides are 'neap tides', when the Sun and Moon are at right angles; here the pulling effect is much smaller and the high and low water marks are very close together.

Tides have such a strong influence on coastal wildlife that beaches are divided into a series of horizontal zones, depending on the extent to which they are covered by water during the tidal cycle. The zones range from the area which is continually covered with water, just below the lowest water mark, up to the splash zone at the top of the beach, which is normally wetted only by sea spray.

All shore wildlife has to adapt to the fluctuating tide levels, the constant splashing of salty water and occa-

Thrift (Armeria maritima) right On the Welsh coast.
Rock samphire (Crithmum maritimum) opposite top It is growing with sea spurrey (Spergularia marina) on an English cliff.

sional fresh water in the form of rain, variations in the concentration of salt in the water, battering by waves, strong winds and the many other rigours of their habitat. The plants and animals cope either by physically resisting them or by avoiding them – for example, by growing in or moving into cool, dark, moist crevices.

However, some coastland habitats are virtually impossible to colonize. On shingle or pebble beaches, the most inhospitable of all marine habitats, the stones are always rubbing together with the action of the waves and any form of life that tries to gain a foothold is quickly rubbed off. On really steep beaches the waves tend to be fiercer and scour more than on gently sloping beaches, giving a similar effect. Sand dunes also shift and provide little or no food or shelter until a few plants have managed to gain a foothold. Even the individual zones on an otherwise suitable beach have such different, and often difficult, features that no single plant or animal can be suitably adapted to living in all of them. Those that can endure the greatest drying (and other constantly changing conditions) tend to live at the top of the shore, while those that require greater stability are found closer to the sea.

All coastland inhabitants tend to be imaginative colonizers. Many seabirds, for instance, have specially designed eggs that will not roll off the ledges on their nesting cliffs; on exposed sandy beaches various invertebrates dig secure homes underneath the sand, their presence betrayed only by tiny feeding casts (worm-like squiggles of excreted sand) or depressions on the surface; others have learnt to hide under rocks, plants or larger animals for protection from the heat of the sun.

Most plants along the shoreline are algae – commonly called seaweeds. Entirely different from land plants, they have no stems, leaves, flowers or fruits. But they have the same basic requirements as their terrestrial counterparts, needing sunlight to photosynthesize and absorbing all their nutrients from the sea.

Seaweeds tend to be the first colonizers on shorelines, and in much the same way marram grass and sand couch grass are the plant pioneers on wind-blown sand dunes. Once they have gained a foothold other species can shelter underneath them, and gradually complex communities begin to take root.

Few animals actually eat seaweeds. Some coastal inhabitants are carnivores and some are scavengers but many are filter feeders, straining their food from the abundant tiny life floating in the sea water. Some species prefer to eat vast amounts of sand, in order to devour the tiny animals that live in the minute quantities of water surrounding each grain.

From the shoreline, the seabed slopes down a gentle gradient known as the continental shelf to a depth of about 150 metres (490 feet). The width of the shelf varies but beyond it the sea bottom falls away sharply to the deep ocean floor. Animals such as jellyfish and octopuses, which normally live in these deeper waters, often come in with the tides and are usually unable to feed; they find themselves stranded in rock pools or on beaches, and if they are not dried out by the sun they have to wait patiently for the next tide in order to escape.

Man has comparatively little use for coastlands, compared with the extent to which other habitats are being utilized and destroyed. But there are many man-made hazards to coastland wildlife, including recreation (causing such problems as the residential development of beautiful coastlines and the disturbance of nesting species), pollution from oil spills, sewage disposal and waste-dumping, and the collection of sea urchins, shells and other animals for the souvenir trade.

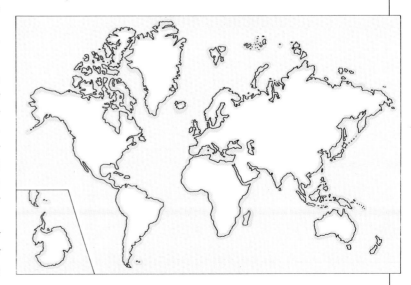

The world's coastlands above
The map shows the vast extent of the coastline habitats.

123

Rock pools

At first glance rock pools look rather like miniature seas, but unlike real seas they are constantly changing as the tide rises and falls. They are pounded by waves when the tide is in and warmed by the sun when it is out. A rainstorm can dilute the salty sea water, or evaporation may concentrate it. Rock pool animals and plants must be able to cope with these and many other changes throughout their daily lives. Among these adaptable species are seaweeds, mussels, starfish, crabs, shrimps and fish.

The beadlet anemone is a very common rock pool inhabitant, often found clinging to the rock surfaces. When the tide is out it looks like a mere blob of red or green jelly, but when covered with water it quickly unfurls its tentacles in readiness for stinging any passing prey. The beadlet anemone feeds on small animals such as shrimps and fish, shooting out a barbed sting as soon as the victim touches one of its tentacles. The dead or dying animal is then pushed into the anemone's hollow, sac-like body through its mouth – which is the same hole through which waste is ejected.

Starfish (Protoreaster lincki) top right
Beadlet anemone (Actinia equina) below
Acorn barnacles bottom right These specimens are on the carapace of a shore crab.

Brittle stars are well-known inhabitants of permanent rock pools and the rocks, stones and seaweeds of the lower shore. Closely related to starfish, which are in fact their greatest enemies, millions of brittle stars sometimes form packed blankets along the shore or seabed. They are famous for their disconcerting habit of shedding an arm if they are handled; this is an adaptation for escaping from predators or in case an arm gets trapped underneath a stone being rolled around by the tide. The lost arm of the brittle star quickly regrows.

Many rock pools are covered with barnacles, which cling on to rock faces as if welded with super-glue. Equally at home attached to marine animals like turtles and whales as they are attached to rocks, barnacles appear lifeless when not covered by water. But as soon as the tide rises and the rock pool fills a trapdoor opens at the top of the shell and a little feathery arm appears to wave about and catch suspended food particles in the water. Although they look rather like snails, barnacles are crustaceans related to crabs and shrimps.

Househunter

Only the front half of a hermit crab is covered with hard shell, so this crustacean has to protect the soft rear parts of its body by hiding in the discarded shells of other animals such as periwinkles and dog whelks. It will even set up home in a small plastic cup if it is the right size. When the crab grows too big it simply goes out house-hunting, inspecting likely looking shells very carefully to make sure they are exactly right.

Other animals often take advantage of the hermit crab's unlikely home. Sometimes sea anemones attach themselves to the shell, in order to hitch rides and grab any pieces of food that the crab stirs up as it feeds and wanders around. In return, the anemone protects the crab from its enemies with its stinging cells. Barnacles, sponges, tubeworms and other creatures also fix themselves to the outside, while ragworms often share the inside of the shell, keeping it clean and tidy by eating leftovers from the crab's meals.

In its early larval stages the hermit crab is quite unlike the adults – so much so that the larvae were once believed to be an entirely separate species. They are planktonic (free-floating) animals and since they get carried around by the water movements this provides the crabs with a good way of dispersing to new habitats. On arrival they quickly change into adults, a transformation as dramatic as the change from caterpillars to butterflies.

Hermit crab *above* With plastic shell.

Sandhoppers

Sandhoppers are sea creatures that have adapted to living on land. They are able to obtain all the moisture they need by burrowing into wet sand or under cast-up seaweed. Only when it gets dark, and the air is cooler, do these tiny relatives of crabs and shrimps venture outside. They swarm out in enormous numbers – in some places there may be as many as 25,000 per square metre (21,000 per square yard) and begin feeding on a variety of vegetable and animal matter, including dead seaweed and the bodies of dead fish, jellyfish and other animals.

If uncovered during the daytime, sandhoppers will jump wildly in all directions. Such unpredictable behaviour is important in defence, since a likely predator is unable to anticipate which way the animals will go and so has difficulty making a kill.

Sandhoppers *(Orchestia ammarella) left*

125

SEA EAGLE

DATA

ENDANGERED

SPECIES
Sea eagle
(*Haliaeetus albicilla*)

CLASSIFICATION
Falconiformes (eagles, hawks,
falcons and vultures)

HABITAT
Rocky coasts, large lakes and
rivers in wild country

SIZE
Up to 90cm (3ft) in length

FOOD
Mainly fish and seabirds; also
carrion, especially in winter

Although sea eagles are often found along rocky coasts, in many parts of their range they prefer to live around wild lakes and rivers. In these areas these birds are commonly called white-tailed eagles since they do not really deserve their other name.

Europe's largest eagle, the sea eagle belongs to the same group of birds as the better-known African fish eagle and the national emblem of the USA, the bald eagle. It is a massive bird with a ponderous, almost heron-like flight, that has disappeared from many parts of its former range. It was fairly common in the early part of the last century, but so many have been poisoned and shot by farmers who (probably wrongly) suspected them of preying on lambs that it has

become extinct in many countries. As the populations dwindled elsewhere egg collectors became interested and the numbers declined even further. Norway is now the last remaining stronghold for the species.

Other birds of prey that prefer a fish diet are the African fish eagle and the osprey. There is an interesting evolutionary discussion between scientists as to whether the first birds of prey were in fact fish-eaters like the sea eagle, with the mammal-eating species evolving from them and eventually becoming more common. This is because some fossil evidence suggests that birds of prey are related to herons, which prey on fish and other water creatures.

Sea eagles eat a considerable amount of carrion (including lambs that have died naturally), but their main prey is fish and seabirds. Not all seabirds are taken; experienced adult eagles always avoid fulmars because they squirt an oily liquid from their mouths which seriously damages the attacker's plumage, impairing both flying ability and insulation.

Most fishing is done by gliding low over the water, though sea eagles occasionally plunge from a great height like an osprey. It is not unusual for them to seize fish which are too big to be lifted clear of the water at the first attempt. When this happens, the eagle floats on the surface, usually with wings spread, before laboriously flapping in the water and having another go at taking off.

When fully grown the female bird is up to 90 centimetres (three feet) in length; the male is slightly smaller. The plumage of both sexes of the sea eagle is dull brown except for the distinctive white tail feathers. Their cry is a distinctive 'kri . . . kri . . . kri'.

The sea eagle's airborne courtship display involves some impressive aerobatics, although it is less spectacular than in many other eagles. The male dives at the female and she rolls over, locking talons with him; together they hurtle downwards, spinning like a cartwheel and unlocking just before they reach the sea or ground. Often a pair has several nests in their breeding area and prefer to refurbish one of these rather than build a new one. Two eggs are normally laid and the young chicks make their first tentative flights when about three months old, although they remain near the nest and are dependent on their parents for a further two months.

PUFFIN

Puffins belong to a family of birds known as auks, which are the Northern Hemisphere equivalents of penguins. Unlike penguins, however, all auks alive today can fly. Only the great auk (*Pinguinus impennis*), which became extinct in 1844, was flightless.

Puffins use their wings both in the air and to 'fly' underwater. They are equally at home diving in rough seas and launching themselves off the top of cliffs. They even bounce undeterred off waves as they take off from the water.

Although they winter at sea, and occasionally appear on inland water during severe storms, puffins visit grassy clifftops and islands to nest. They gather offshore in early spring before flying in large flocks to the coastlands, using their webbed feet spread as 'air-brakes' when landing. Colonies sometimes number thousands of pairs, each with its own self-made nesting burrow or converted rabbit hole in which to lay the eggs.

Both parents share incubation and they feed the single chick together for about six weeks. It was once believed that they deserted the young bird after this time and starved it out of the burrow, but they are now thought to continue feeding until the day it leaves. Travelling after dark, to avoid hungry gulls and skuas, the chick makes its way to sea and does not return to the colony for a further two or three years. The birds are at least four or five years old before they return to breed themselves.

Puffins have suffered in many parts of the world because both eggs and adult birds are taken by people for food. And since they spend so much time at sea they are also particularly vulnerable to oiling. They are often inadvertently caught and drowned in fishing nets. These days competition with man for fish is increasingly a problem, now that larger species such as cod have been virtually fished out and many of the puffin's main food items (such as capelin, sand eels and sprats) are being caught on a commercial basis instead.

Puffins are real fishing experts, able to catch ten or more small fish in succession without having to swallow; their bills are so enormous that they can carry all the fish crosswise. The bills are also important in pair formation and courtship, which is why they are so large and colourful in summer.

DATA

SPECIES
Puffin
(*Fratercula arctica*)

CLASSIFICATION
Charadriiformes (waders, auks, gulls and others)

DISTRIBUTION
Breeds on islands and coasts bordering North Atlantic and Arctic Oceans; winters at sea

HABITAT
Nests on grassy clifftops and islands

SIZE
30cm (1ft)

FOOD
Small fish such as sand eels, capelin, whiting and sprats

SEA OTTER

D A T A	
SPECIES	
Sea otter	
(*Enhydra lutris*)	
CLASSIFICATION	
Carnivora (carnivores)	
DISTRIBUTION	
North Pacific, in western North America and eastern coasts of USSR	
HABITAT	
Rocky coasts and kelp beds	
SIZE	
Up to 1.2 metres (4ft) plus tail of about 30cm (1ft)	
FOOD	
Crabs, mussels, snails, fish, squid and sea urchins	

Most otters are so full of energy that they hardly seem to keep still. The sea otter, which spends almost its entire life in the sea, is no exception. Although sea otters rarely venture more than a kilometre from the shore they only haul out of the water during very severe storms or, very occasionally, to rest on a rock. Although not very good at walking on land otters are expert swimmers, paddling along with their hind limbs and using their tails as oars. If they are in a hurry sea otters also use seal-like undulations of their bodies which enable them to swim at speeds of up to nine kilometres (six miles) per hour.

Sea otters are active only during the day, when they regularly dive to depths of 20 metres (65 feet) or even 40 metres (130 feet) in the northern parts of their range, to search for crabs, mussels, sea urchins, snails, squid, fish and other food. Dives average slightly over one minute in length, during which the otters catch prey with their front paws rather than their mouths; then they surface and eat while floating on their backs in the water. Using their chests as tables, they often use rocks to smash the hard shells of some animals.

After they have finished eating the otters roll over in the water to wash bits of shell or leftover food from their fur. Sea otters need enormous amounts of food to survive and may eat as much as 10 kilograms (22 pounds) a day. At night they tie themselves to pieces of kelp (a kind of large seaweed) to avoid drifting away with the currents. They sometimes sleep with their paws over their eyes.

The sea otter is the only otter that does not come ashore to give birth. The single offspring has to be carried around on its mother's chest, while she swims on her back, until it is able to dive alone at six or seven weeks old. The two stay together for up to eight months, during which time the mother spends a great deal of time playing with the young otter and teaching it to dive for food.

Sea otter fur is extremely valuable and the animals were exploited extensively for over two centuries. By 1911 their numbers had dropped to a little over 1,000 but, thanks to complete protection since then, the population has fully recovered and at least 100,000 are alive today.

There is, however, one relatively new but potentially disastrous threat to their survival – oil pollution. Unlike whales and seals, sea otters do not have a special layer of fat under their skin to keep warm. Instead, they rely on a layer of air trapped among the long, soft fibres of their fur. If the hair is damaged – for example, by oil – the insulating properties are lost and the otter can easily die of cold. An oil spill could kill enormous numbers of the animals and perhaps wipe out entire otter populations.

SEA URCHIN

Sea urchins are strange-looking animals related to starfish and sea cucumbers. Found along coastlines, on the seabed or buried in the sand, they creep around camouflaged with scraps of seaweed and shells in search of plants and animals to eat. They occur in all the world's seas. Most species are found between the low tide mark and 200 metres (650 feet) down, but some live much deeper; one species has been found at a depth of over seven kilometres (four miles).

They also have shells, or 'tests', of their own. Made of calcium carbonate, these are usually between six and twelve centimetres (2½ to 5 inches) in diameter – but some Indo-Pacific sea urchins can reach diameters of nearly 36 centimetres (14 inches). All the soft parts of their bodies are protected inside but, even so, young ones are often eaten by fish, seals, sea otters and other animals. The adults are therefore better protected with a formidable array of tough spines on top of their shells. In fact, the scientific name for sea urchin, Echinus, is very appropriate because it is actually Greek for 'like a hedgehog'. The spines are mobile, mounted on small knobs on the outside of the shells, and enable the animals to move around. When the urchins die, the spines fall off and, unfortunately, in many tourist areas, their spineless shells have become popular as souvenirs. It is a sad fact that many living animals are taken from the wild and killed for this purpose too.

Sea urchins also have special 'feet', known as tube feet, which pop through holes all over the shell. Tube feet are really muscular tubes filled with liquid. They have very strong suction and enable the urchins to climb vertical rock faces and kelp stems, where they can scrape off seaweed and tiny animals with their strong teeth.

Sea urchins feed with a complicated apparatus known as 'Aristotle's lantern', which consists of five chisel-shaped teeth, worked by a very complex arrangement of muscles.

Some sea urchins use their spines and tube feet in combination to plough through soft sand, picking food from the sand particles with their pedicellariae – specialized spines with tiny jaws on the ends. Others burrow up to twice their diameter into the sand, extending extra-long tube feet to the surface to breath. There are a few species which extend this burrowing to rock, using their spines and teeth to rasp a hole. In danger they will drop to the bottom of their hole and wedge themselves in solidly with their spines. The power of these sea urchins is incredible. There is a Californian species of urchin which in 20 years managed to drill itself ten millimetres (three-eighths of an inch) into a solid steel underwater girder by the same method.

They usually breed early in the spring. The males and females release their sperm and eggs into the water at the same time and the fertilized eggs develop into minute larvae. Along with many other minute creatures these little animals are known as plankton. They often travel enormous distances with the sea's currents and have no control over their destinations, so many die before they become adults. However, they are the sea urchins' only chance of colonizing new areas. Although the adults can travel several miles on their spines and tube feet, they generally spend most of their time in one small area.

DATA	
SPECIES	Edible sea urchin (*Echinus esculentus*)
CLASSIFICATION	Echinodermata (starfish, sea cucumbers, brittle stars and others)
DISTRIBUTION	European coasts
HABITAT	From low shore to depths of 50 metres (160ft) or more
SIZE	10cm (4in) in diameter
FOOD	Plants and dead animals

GREY WHALE

DATA

ENDANGERED

SPECIES
Grey whale
(*Eschrichtius robustus*)

CLASSIFICATION
Cetacea (whales, dolphins and
porpoises)

DISTRIBUTION
North Pacific and southern
Arctic Oceans

HABITAT
Coastal waters

SIZE
Length averages 12.2 metres
(40ft) for males, 12.8 metres
(42ft) for females; weight up to
38 tonnes

FOOD
Bottom-living invertebrates
such as crustaceans, molluscs
and bristle worms

Grey whales were hunted so intensively in recent centuries that at one time it was thought they were extinct. They did disappear altogether from the North Atlantic, but since 1946 have been fully protected elsewhere. There are two surviving populations, in the western and eastern Pacific. The western population, which lives in the Okhotsk Sea and the Sea of Japan, is still critically endangered and may even be extinct, but the eastern population is slowly recovering and now stands at between 15,000 and 17,000 individuals.

The grey whale is one of the few large whales to prefer shallow, coastal waters. Measuring up to 15 metres (nearly 50 feet) in length and weighing as much as 38 tonnes (the tongue alone weighs some 1.4 tonnes) it is an easy animal to see. Since it sometimes ventures less than a kilometre from the shore, it has become one of the best known of all whales. Even its spout of vapour breathed out after a dive is distinctive, rising over four metres (13 feet) into the air and audible from more than a kilometre away.

Grey whales are particularly famous for their 20,000-kilometre (12,500-mile) round-trip migrations, between the Arctic Ocean or Alaska and Baja California, in Mexico. These are designed to help the whales exploit the tremendous abundance of invertebrate food to be found in the Arctic seas during the summer. They feed by ploughing along the sea bed, stirring up the sediment and filtering out a variety of invertebrates such as small shrimps and shellfish as well as the occasional fish or piece of seaweed which gets in the way.

In late September and early October the whales begin moving south towards their breeding grounds, at a rate of about 180 kilometres (110 miles) a day. The journey takes them up to two months, during which time they eat virtually nothing. On arrival in Mexico, they continue fasting and begin a very complex and elaborate form of courtship. The pregnancy lasts exactly 12 months, which means that the single calf is born in the same lagoons the following year. The birth takes place in water less than ten metres (33 feet) deep.

When the infants are about two months old, which is usually in late February or early March, they accompany their mothers on the return journey northwards for the first time, with the males close behind. This journey often takes twice as long as the southward one, because of opposing currents and the presence of the young calves. In the Arctic they stay with their mothers as they feed, before returning once again to their winter homes in the south.

Grey whales seem indifferent to ships and boats – they have even been known to lay their heads on the sides of small boats to be patted. Unfortunately disturbance in their breeding grounds by well-meaning tourists, as well as industrialization and commercial shipping on their migration routes, are the main problems they face today.

LOGGERHEAD TURTLE

Superbly adapted to life in the sea, and able to dive and swim underwater for long periods with their legs modified into powerful flippers, sea turtles are actually descended from land-living tortoises. In fact their eggs retain this terrestrial ancestry and can only develop and hatch on land, so every year the female sea turtles have to leave the safety of their deep-water homes and haul themselves out onto sandy beaches to lay.

Loggerhead turtles are one of the larger species of sea turtles and are easily distinguished by their disproportionately large heads. They lay their eggs in late spring and summer. The males never leave the water but the females slowly clamber up to the back of the beach, usually under cover of darkness. Each digs a large hole in the sand with her hind flippers and lays inside 100 or more white, round eggs, each about the size of a ping-pong ball. Once the eggs are safely covered with sand the turtles return to the sea. The average female repeats this two or three times a season.

When the five-centimetre (two-inch) long hatchlings clamber out of the sand they scurry across the beach to the sea as quickly as they can. Many are caught on the way and eaten by gulls, skuas and other predators.

There are relatively few suitable undisturbed, sandy nesting beaches for sea turtles these days. Although loggerheads are found worldwide, in temperate and subtropical waters from the USA and Greece to Mozambique and Australia, they have been squeezed out of many of their former nesting sites by people. Hotels are built at the backs of their beaches and tourists trample over the nests. Many of the turtles that survive are killed for their meat or shells, which are fashioned into curios and sold to tourists. Large numbers are also drowned or battered to death after being caught accidentally in fishing nets; up to 4,000 loggerheads die in this way every year off the south-east coast of the USA.

We are not certain how many loggerhead turtles are left, since they are so difficult to count. There are still some sizeable colonies – one in particular, in Oman, has 30,000 females nesting annually – but most populations have declined significantly. Sadly, the same is true of other sea turtles. Six of the world's seven species are endangered and the seventh is very rare.

DATA
ENDANGERED
SPECIES
Loggerhead turtle (*Caretta caretta*)
CLASSIFICATION
Chelonia (turtles, terrapins and tortoises)
DISTRIBUTION
Temperate and subtropical waters worldwide
HABITAT
Coastal waters, nests on undisturbed sandy beaches
SIZE
Body up to 1 metre (40in) in length; weight up to 160kg (350lb)
FOOD
Bottom-living invertebrates such as molluscs, sponges and shellfish in shallows; salps and jellyfish in deeper water

OCEANS

The seas and oceans cover more than seven-tenths of our planet's surface. Divided into four main areas by the continents and islands they are the Atlantic, Arctic, Pacific and Indian Oceans, together with their adjacent seas such as the North Sea and the Mediterranean.

Our oceans are in a constant state of flux. The pull of the Sun and Moon on the Earth, combined with the relative movements of these bodies, causes a daily rise and fall of the water levels, known as tides. There are also internal water movements, or currents.

Ocean depths vary considerably throughout the world. All land is fringed by an underwater 'continental shelf' which is gently-sloping and extends from the tideline to a depth of about 200 metres (650 feet). The shelf can be anything from 16 to 1,250 kilometres (10 to 780 miles) wide. Beyond this shelf the land slopes more steeply to the ocean floor, in places many kilometres below sea level, where there is a hidden underwater world of mountains, canyons, valleys and cliffs. Mount Everest would fit into the deepest part of the ocean and still be covered by about a kilometre (over half a mile) of water.

Deep down near the ocean floor it is permanently dark. The waters are near freezing and the pressures are immense. Even so, an incredible variety of creatures live there, surviving on one another and a meagre diet of dead plants and animals which drift down from surface waters above. Among these deep-sea animals are angler fish, gulper fish and more than 2,000 other species, some with their own light-producing organs to communicate with one another and to attract prey. Indeed there are probably many more weird and wonderful creatures but the deepest oceans are, as yet, largely unexplored so we do not know exactly what is down there.

Most marine animals and plants live in the surface 150 metres (500 feet) or so. This is the warmest layer and the only part of the sea which sunlight is able to penetrate. It is possible to tell how much life there is in this region by its colour. Sea which is blue generally has little marine life. But green sea is rich – it is actually murky with minute floating plants known as phytoplankton, and so contains the food on which all species ultimately rely.

The sea may seem a hostile and dangerous environment to land animals such as ourselves but it offers many advantages. Indeed, it is in the sea that life first appeared some 3,500 million years ago. It has much smaller temperature ranges than land, and the cold and salty water is much more buoyant than air so it is not essential for animals to have strong skeletons and muscles to support their bodies. This buoyancy has allowed the appearance of the largest animal that has ever lived, the blue whale (Balaenoptera musculus). This mammal weighs about 100 tonnes and could never have lived on land simply because such a colossal body could not support itself in air.

Some animals, mostly in shallow and clear tropical seas, are comparatively tiny themselves but build their own enormous living habitats. Small, soft-bodied creatures called polyps, related to jellyfish and sea anemones, are able to take the chemical calcium carbonate from sea water and use it to make hard limestone skeletons, called coral. A type of algae which gets its nutrients from the polyps then helps to build up 'coral reefs' by cementing the different structures together.

Together they form what can only be described as tropical forests under the sea. Reefs 2,000 kilometres (1,250 miles) or longer, like the Great Barrier Reef in Australia, are a blaze of colour and house a wealth of wildlife – fish, sea slugs, starfish, jellyfish and sea cucumbers among them. But the reefs are threatened by polution and digging, resulting in a loss of species and of the natural barriers that protect nearby land.

Sadly coral reef destruction is just one of many irresponsible human activities affecting the seas. Habitat destruction on land is a problem but the oceans are even less well cared for. Two-thirds of the world's marine fisheries depend directly on environments such as estuaries, mangrove swamps and salt marshes, which are crucial for spawning and as fish nurseries. Yet as they are being degraded or destroyed outright. Countries around the world are taking too many fish from the waters under their own national jurisdiction, or haphazardly ruining them with pollution.

The world's most valuable animals – shrimps, cod and herring – live in the sea. These alone are worth many thousands of millions of pounds a year. But too many are being taken for the populations to maintain their numbers. Over-exploitation in the past means that catches of many species are now a fraction of what they could have been with proper management.

Not only fish and shrimps are under pressure. Seals, whales and other marine animals are also being over-exploited. The plight of the world's whales is now so acute that a moratorium on whale-hunting is considered essential by scientists around the world. But this is not being adhered to by a number of whaling nations.

Pollution is also a serious danger threatening the world's oceans. Oil is released in vast quantities into the sea through accidents on oil rigs and after the grounding and routine cleaning of tankers. The oil forms 'slicks' on the surface of the sea which are death traps to seabirds and other animals. Sewage and poisonous metals, either dumped into the sea directly or entering via rivers, are other major pollutants. Too much sewage, for example, causes an excess of nutrients that permits huge growths of algae, which then use up all the oxygen and block the light. Poisonous metals eaten by small sea creatures can ultimately cause the death of large marine mammals which feed on them.

We do not know how many marine species are plunging to extinction in these ways. We know so little about the oceans that there are probably many creatures which have not yet been discovered or named, many which may prove to be invaluable to human survival and comfort in the future. Yet the destruction continues.

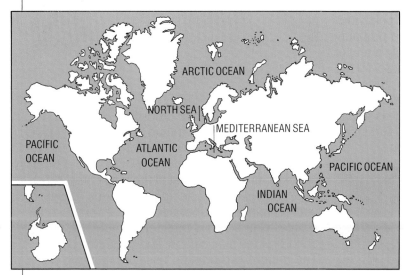

High surf opposite top
On the north shore of Oahu, Hawaii.
Seaweed (Ecklonia radiata)
opposite bottom

The world's oceans above
Oceans and seas cover over seventy per cent of the world's surface.

Hidden dangers

Sand tiger shark (*Odontapsis taurus*) *above*

Many animals are potentially dangerous to people, though most are harmless unless provoked by our own idiotic or insensitive behaviour. There are some hidden dangers beneath the ocean waves, because we cannot always be aware of the animals' presence in order to behave sensibly, but even these tend to be over-exaggerated.

Understandably, sharks are among the most feared animals in the world. A shark attack is usually a silent and secret affair – it is sudden, swift, unexpected, and often fatal. Yet two of the biggest sharks, the 15-metre (50-feet) whale shark *Rhincodon typus* and the 10-metre (32-feet) basking shark *Cetorhinus maximus*, are both harmless plankton strainers. Even many of the so-called man-eaters will swim past people with no more than a passing interest.

The greatest problem is that shark behaviour is unpredictable and most attacks are probably attributable to rogue individuals of only a few of the 250 or so shark species. Among these are the great white (*Carcharodon carcharias*), the hammerhead (*Sphyrna mokkaran*) and the sand tiger shark (*Odontaspis taurus*); for some reason this last species is considered harmless in American waters but highly dangerous elsewhere.

Scorpionfish *right*
Olive sea snake (*Aipysurus laevis*) *below*

The scorpionfish, potentially a highly poisonous animal, tends not to swim around looking for people to injure or eat. But the spines on its body are attached to extremely potent venom glands, and since these fish are remarkably well camouflaged they are easy to tread on. Also called zebrafish, lionfish, dragonfish, firefish, devilfish, turkeyfish or featherfish, the scorpionfish is a bizarre and beautiful sight as it lies motionless, waiting for a crab, shrimp or another fish to pass nearby. But it does not wait to be stepped on by intruders, preferring constantly to manoeuvre its spines in the direction of any unsuspecting foot. There have been few human deaths from scorpionfish stings but there are many records (from all the world's seas and oceans) of the agonizing injuries inflicted by them.

Sea snakes, which are mostly residents of the western Pacific and Indian Oceans, are also highly venomous. There are approximately 50 species and at least one is known to have venom 50 times as potent as that of the king cobra. Most bites are linked with fishing operations (they are often caught in nets) or because they are coastal reptiles with a strong preference for beach areas and coral reefs, inevitably frequented by people. The venom is normally used for hunting fish, which are obtained by striking in much the same way as any other poisonous snake strikes; since sea snakes are not generally aggressive animals, bites on humans are rarer than might otherwise be expected.

Seahorses

Surprising though it may seem, seahorses are fish and are believed to be related to sticklebacks. However, they are very weak swimmers and the only way they avoid being carried along with even the slightest ocean current is by clinging to the nearest seaweed with their prehensile tails. In still water they are able to hold their bodies upright and use what in other fish is the dorsal fin as a kind of propeller. This looks very impressive, enabling the animals to move elegantly through the coral reefs and weed forests in which they live, but is not particularly efficient. The propeller system does enable them to swim vertically up and down, though – allowing some manoeuvres which most other fish cannot manage.

The male seahorse has a 'false stomach', or pouch, underneath his tail which is used to store the eggs. The female lays these inside, and when they hatch the pouch is used to carry the young. Many people, seeing this extraordinary sight, mistakenly believe that the male seahorses, with their full pouches, are pregnant.

Seahorse (*Hippocampus kuda*) *left*
This specimen is a male, showing the pouch where the eggs are stored.

The ocean food supply

Krill are the tiny shrimp-like creatures on which most other ocean animals ultimately depend. There are approximately 90 species; an individual is on average about three centimetres (just over an inch) long, though some, such as the Antarctic krill (*Euphasia superba*), can be twice this size. They live in the open seas, some near the surface and others at depths of 2,000 metres (6,500 feet) or more.

Many types of krill live in enormous shoals, often in such quantities that they colour the sea pink. A single shoal may occupy an enormous volume under the water, packed together at densities of up to 63,000 individuals per cubic metre. They are often so vast and dense that they can be detected by remote-sensing satellites.

Krill is the major component of zooplankton, the tiny floating animals on which myriads of seabirds, seals and whales feed. A blue whale may eat up to three or more tonnes of krill in one go as it swims through a shoal with its huge mouth open. As it takes gulps of 'krill soup' it uses its tongue (which itself may weigh as much as an elephant) to squeeze the water out through special sieve plates; then it swallows the animals left behind.

Antarctic krill (*Euphausia superba*) above

MINKE WHALE

DATA

SPECIES
Minke whale
(*Balaenoptera acutorostrata*)

CLASSIFICATION
Cetacea (whales, dolphins and porpoises)

DISTRIBUTION
Worldwide, though rare in the tropics

HABITAT
Prefers shallow water; sometimes occurs in estuaries, inland seas and rivers

SIZE
Average length about 8 metres (26ft); weight 6 to 7 tonnes (tons)

FOOD
Fish, squid and plankton

The minke whale is more likely to be seen at close quarters than most other species of whale. It favours shallow water near the coast and sometimes ventures into inland seas and rivers. It also seems to be curious about ships and boats and often approaches to have a closer look, easily keeping pace with vessels moving at 16 knots or more. This is an unfortunate habit because it makes minkes easy targets for whalers. Indeed, enormous numbers have been killed in past decades. No-one knows how many are left: some say as many as 250,000, others put the figure nearer 100,000 or even lower. But there is no doubt that numbers have decreased considerably since the early 1940s, when large-scale exploitation began. The north-east Atlantic population, for example, which has been hunted mostly by Norway, now stands at a quarter or less of its original size, with more than 100,000 having been killed in less than half a century.

Despite drastic declines in their numbers minke whales, or piked whales as they are also known, are still widely distributed throughout the world. They even occur right up to the edges of Antarctica and Greenland, though usually only the bigger, older animals venture this far.

Minkes usually mate in the winter, in temperate or tropical waters on either side of the equator, where they live singly or in pairs. But as summer approaches they migrate much farther north or south to feed on the rich shoals of plankton which can be found in the Arctic and Antarctic at this time of year. Many hundreds often gather together in particularly favoured feeding grounds. After three or four months spent gorging themselves they migrate back to the same waters in which they mated some 10 or 11 months earlier, to give birth. The newborn minke, which at about three metres (10 feet) long is over a third the length of its parents, is suckled for about six months. All the members then feed on fish (especially cod, anchovy and herring) and squid before heading back to the polar waters to feed on plankton once again.

MEDITERRANEAN MONK SEAL

Over two thousand years ago the Greek philosopher Aristotle gave an accurate account of a seal which used to be a very common inhabitant of the Mediterranean. Well known to the local people for centuries, it used to live in small herds and was a regular sight during the daytime on the sandy beaches around Greece, Turkey, Italy, Morocco and a number of other countries.

Today that same seal, known as the Mediterranean monk seal, can still be seen. But tourists have frightened it away from its original homes on the sandy beaches; hunters and fishermen have killed large numbers for their skins, oil and meat, or to stop them eating too many fish; and pollution has also taken its toll. The result is that the Mediterranean monk seal population has dropped to only 500, or even lower. These survivors are so scarce that they have become widely scattered, living alone or in small groups of twos and threes. They are effectively restricted to small islands which are uninhabited by man because of a lack of fresh water and to cliff-bound rocky beaches in only a few parts of their original range.

Very sensitive to disturbance of any kind, even in these apparently safe locations, the monk seals have been forced to adapt to more nocturnal habits and prefer caves with underwater entrances to avoid having to haul themselves about in the open. The young, too, are nearly always born in the shelter of these sea caves, usually in September and October, instead of on rocks and beaches as before.

Today the Mediterranean monk seal is protected in most of the countries in which it still occurs, but is still regarded as a pest by many fishermen. Its only hope of survival is the establishment of a sufficient number of well-guarded nature reserves and the strict enforcement elsewhere of protective measures to prevent individuals being killed. Without such conservation efforts, the species can scarcely be expected to survive much longer. For frighteningly similar reasons a closely related species, the Caribbean monk seal (*Monachus tropicalis*), is already thought to be extinct and a third, the Hawaiian monk seal (*Monachus schauinslandi*), is heading for the same fate.

DATA

ENDANGERED
SPECIES
Mediterranean monk seal (*Monachus monachus*)
CLASSIFICATION
Pinnipedia (seals, sealions and walrus)
DISTRIBUTION
Small areas of the Mediterranean, a few around the Canary Islands and in part of the western Black Sea
HABITAT
Breeds on uninhabited islands and cliff-bound rocky coasts
SIZE
Up to 3 metres (10ft) long; weighs up to 400kg (880lb)
FOOD
Fish and octopus

BOTTLE-NOSED DOLPHIN

The bottle-nosed dolphin (*Tursiops truncatus*), found in coastal waters in many parts of the world, is one of 32 species of dolphin. All are generally friendly animals – even the larger members of the group such as killer whales (*Orcinus orca*) do not harm people if unprovoked – and their intelligence is probably unequalled except by the primates, and perhaps only by man. Dolphins also have a highly developed social organization, sometimes living in coordinated schools of 1,000 individuals or more. Unfortunately for populations in the wild their intelligence and ability to execute graceful and spectacular leaps (which are often done for fun in the wild) have made them highly sought after by zoos and oceanaria, and large numbers are killed because of the presumed conflicts with fisheries.

PARROTFISH

D A T A

SPECIES
Parrotfish
(*Scaridae*)

CLASSIFICATION
Perciformes (perches, butterfly-
fish, barracudas and many
others)

DISTRIBUTION
Tropical and subtropical seas

HABITAT
Coral reefs

SIZE
May exceed 2 metres (nearly
7ft) though most species
considerably smaller

FOOD
Algae and coral

There are 80 species of parrotfish living along the steep ridges of coral reefs in tropical and subtropical seas. They are almost all colourful fish and their hues and patterns vary greatly from individual to individual. Not only do the males and females look completely different, but many of the young fish go through two or three colour phases before they become adults. This has led to great confusion in the past and scientists once thought that there were over 350 different species. Even now intensive field studies may result in further reductions and there are still several names for different-coloured fish of the same species.

Parrotfish are so called because their 'beaks' are very similar to those of parrots. The 'beaks' are formed by their teeth, which have fused together at the front and are used for scraping at coral. As the tide goes in and out the fish move up and down the reefs, feeding on coral and algae. After scraping away the hard material with their beaks, they use another set of teeth farther back in their mouths, and specially designed for grinding, to crush the coral and algae mixture. The hard pieces are then spat out and any indigestible sand is excreted.

This type of feeding causes a considerable amount of erosion, and parrotfish are the main cause of reducing coral reefs to sand. They produce the countless tonnes of white sand which rings so many tropical islands and accumulates on beaches on the mainland.

Fortunately, damage to the reefs is not usually very serious. The tiny coral animals, or polyps, quickly regrow on bare spots in most areas. But even if part of the reef collapses after parrotfish have been at work, many new hiding places (in the form of gaps and tunnels) are provided for an enormous variety of fish and other sea creatures. So the reef can support even more wildlife.

The parrotfish themselves often hide in these nooks and crannies but they also have an additional form of protection. They secrete a shiny layer of jelly-like substance around themselves every night, which forms a protective sleeping cocoon. It takes about half an hour to build – and almost the same time again is needed to break out of it in the morning. Although not all species of parrotfish make such balloon-like 'pyjamas' they are thought to be an effective means of protection against forays by nocturnal predators such as moray eels.

MARLIN

The marlin belongs to a family of spectacular ocean fish known as the billfish. There are about 30 species in the group, widely distributed in the Atlantic, Indian and Pacific Oceans. They include the sailfish, whose dorsal fin is so large that it really does form a sail several times the height of its own body; also the lance nosed spearfish and the super-streamlined bonito, wahoo and tunny.

Rather like the well-known swordfish, all these billfish have an extended upper jaw which forms a long, spear-like contraption sticking out in front. There are several reports of billfish using these to hole and sink boats of considerable size; whether or not this is true, they certainly are powerful creatures and the 'sword' or 'spear' is a formidable weapon. It is often used in an amazing hunting technique whereby the billfish dashes into a school of smaller fish, such as mackerel, and jerks its head from side to side while swimming through. It then returns and eats the wounded, torn-up prey.

Another characteristic of billfish is their superbly streamlined, torpedo-shaped bodies. Hydrodynamically they are about as close to the ideal speed-swimming shape as it is possible to get. One species in particular, the sailfish, can reach speeds in the water of over 110 kilometres (70 miles) per hour – even quicker than the cheetah, fastest of all land animals.

Marlins are also very fast swimmers. There are four species, all deep-sea fish, at least one of which can reach 80 kilometres (50 miles) per hour. They make great leaps out of the water, which is a spectacular sight considering their size. The blue marlin, for example, can jump extraordinary distances of over 40 metres (130 feet). The largest of the four is the black marlin, from the Indo-Pacific; it can weigh over 700 kilograms (1,500 pounds) and may reach a length of 4.5 metres (14½ feet). Their tremendous size and speed, and the fact that they are terrific and almost tireless fighters, make marlins among the most challenging and popular sports fish in the world.

The striped marlin is an impressive fish. The biggest specimen caught by rod and line weighed in at 190 kilograms (418 pounds) though larger specimens have been recorded. The fourth member of the group, the white marlin, lives in Atlantic waters and the record game-fishing catch is 79 kilograms (174 pounds).

Although a marlin makes a fine 'prize' for the sports fisherman, after being hooked many are simply weighed, tagged with a small metal identity disc and released back into the sea. Using a 'tag and release' programme, if the fish is caught again its movements can be charted. Findings reveal that, for example, some individuals have crossed the entire Pacific, from Australia to the west coast of South America. Even those marlins that are caught and killed can provide useful information. For instance, one particularly large specimen obtained by a research ship had a tuna over one and a half metres (five feet) long in its stomach – swallowed whole. This tuna had two holes in it, almost certainly made by the marlin's spike.

D A T A
SPECIES
Striped marlin
(*Tetrapturus audax*)
CLASSIFICATION
Perciformes (perch, butterflyfish, sunfish, barracuda and others)
DISTRIBUTION
Warmer waters of Pacific Ocean
HABITAT
Deep seas, very occasionally in coastal waters
SIZE
Length up to 3 metres (10ft); weight up to 270 kg (600lb)

HORSESHOE CRAB

DATA

ENDANGERED

SPECIES
Horseshoe crab
(*Tachypleus gigas*)

CLASSIFICATION
Xiphosura (king or horseshoe
crabs)

DISTRIBUTION
East coasts of Asia and USA

HABITAT
Ocean floor at considerable
depths; sandy coasts for breeding in spring

SIZE
Up to 30cm (12in) plus 'tail' of
roughly same length

FOOD
Worms, soft-shelled molluscs
and other small creatures

The horseshoe crab has remained unchanged since it first appeared on earth over 200 million years ago. Unlike any other species alive today, this strange prehistoric animal is not a crab at all; it is probably related to a group of extinct marine creatures known as the trilobites.

Found mostly in the warmer seas of the world, horseshoe crabs breathe through special 'book gills' on their abdomens. These consist of about 150 thin plates arranged like the leaves of a book. The crabs have long spear-like tails which are highly mobile and used for pushing and righting themselves when they are accidentally turned over; however, these ominous-looking structures are never used for defence and the creatures can safely be picked up and carried by them.

There are five species of horseshoe crab, some living off the east coasts of Asia and the others in seas along the North Atlantic coast of the USA. Although common they are rarely seen for most of the year as they spend their time at considerable depths, searching for worms, clams and other small creatures on the ocean bed. However, every spring they migrate towards the coasts to breed and this is when they are easily found. They gather in shallow waters until full moon and then, on three consecutive nights when the tide is high, hundreds of thousands of them emerge from the sea. The males climb onto the females' backs, clinging on with their specially hooked legs, and are dragged to the edge of the water. On favourite beaches a continuous mass of crabs can be seen stretching for kilometres along the coast. The females half bury themselves in the sand and shed up to one thousand eggs, prompting the males to release their sperm.

Hordes of gulls and small wading birds congregate to feed on the fertilized eggs. Those that survive, buried deep in the sand, stay there for a month until it is full moon once more and the high water reaches them. The larvae, only two centimetres (less than one inch) across and as yet minus their parents' armoured plating, escape and swim freely out to sea.

GIANT CLAM

NAUTILUS

In the deep waters of the Pacific and Indian Oceans there lives a strange shelled animal, known as *Nautilus*. There are half a dozen similar species in the group, up to 25 centimetres (10 inches) in diameter, living on the ocean bed at depths of between 50 and 650 metres (160 and 2,100 feet). They are nocturnal predators, usually remaining hidden during the day but coming out at night to feed on crabs and carrion. Their shells consist of several successive chambers – as the animals grow new ones are added – which are filled with gas and fluid. By regulating the amount of fluid (and thereby influencing the gas pressure) the animal can control whether it sinks or floats. It is even able to swim, albeit jerkily, by jet-propulsion using the same technique.

At over one metre (three feet) across and weighing more than a quarter of a tonne, the giant clam is the world's largest mollusc. There are many stories about man-eating giant clams and pearl-divers drowning after getting their legs caught in them. It is certainly true that, although brightly coloured, these clams often have a great deal of coral growing around them and so can be difficult to see; also, the muscles which hold the two shells or 'valves' together are tremendously strong; and the serrated edges of the valves fit perfectly together like giant teeth. But such stories are almost certainly legends since the valves close so slowly that a diver would have to be asleep to get caught. Besides, clams are harmless filter feeders and live solely on microscopic organisms, obtaining them by sucking in water, filtering it and spitting it out minus its microscopic food.

The exposed 'lips' of the giant clam have a number of small single-celled algae living on them. Together, the clam and algae form a living partnership which is of mutual benefit. The algae obtain food substances from the clam and in return provide an additional supply of carbohydrates and oxygen for their host. The relationship is so important that clams are nearly always found in shallow waters where there is enough sunlight for the algae to live.

Giant clams have been used by people for centuries. In the past they have been collected for use as washbasins or baptismal fonts, and more recently as the providers of giant pearls. These are produced by many marine and freshwater 'bivalve' molluscs and consist of the chemical calcium carbonate (with various organic ingredients) which is made by the mollusc in an attempt to encase and kill minute but troublesome parasitic worms. Giant clams have yielded giant pearls weighing six or seven kilograms (15 pounds) and worth hundreds of thousands of pounds.

D A T A
ENDANGERED
SPECIES
Giant clam
(*Tridacna gigas*)
CLASSIFICATION
Mollusca (slugs, snails, mussels, squids and many others)
DISTRIBUTION
Indian and Pacific Oceans
HABITAT
Coral reefs
SIZE
Up to 1 metre (3ft) in length; weighs 200kg (440lb) or more
FOOD
Plankton

WANDERING ALBATROSS

DATA

SPECIES	Wandering albatross (*Diomedia exulans*)
CLASSIFICATION	Procellariiformes (albatrosses, shearwaters and petrels)
DISTRIBUTION	Between Antarctic Circle and Tropic of Capricorn
HABITAT	Open oceans
SIZE	Wingspan of up to 3.5 metres (11⅓ft)
FOOD	Surface fish, squid and offal from ships

Beautiful, graceful and seemingly tireless, the wandering albatross spends many more hours on the wing than at rest on the water. Affectionately nicknamed the 'mollymawk' or 'goony', it is often found hundreds of kilometres from the nearest land. Indeed, the wandering albatross has been a familiar and popular sight among sailors for centuries.

Wandering albatrosses beat their wings as little as possible to conserve energy, but can still glide effortlessly at speeds of up to 100 kilometres (60 miles) per hour. This is one reason why they live in the windiest oceans of the world and are rare, even on passage, elsewhere. The wind causes waves which indirectly provide turbulent uplifts of air, and thereby excellent conditions for gliding. The albatrosses glide downwards to within one metre (yard) of the water, then swing up with the lift to about 20 metres (22 yards), losing speed as they rise before diving down into another glide to regain full speed once again.

There are 14 species of albatrosses; with a wingspan of up to three and a half metres (over 11 feet) the wandering is the largest. At its breeding colonies on oceanic islands there has to be plenty of room for their long wings to be used in elaborate courtship displays. The male arrives first, early in the southern spring, and chooses an open site (often one it has used before) with plenty of wind for taking off and landing. Three to five weeks pass before the female arrives, by which time he has made a basic nest of tussock grass, moss and mud. After a spectacular greeting ceremony of bill-clapping and wing displays, to reintroduce themselves after the long separation since their last mating two years earlier, the pair breed and the female lays her single egg.

The parents take turns to incubate, changing over every week or so for a total of 11 weeks. The chick hatches as winter sets in. It remains in the nest for no less than 10 months (a record among birds) before launching itself into the air and out to sea. Like its parents, the young albatross flies much too fast to be able to pick up food on the wing. It has to settle on the water first before swimming towards its prey, which includes fish, squid and offal from ships. The young bird remains an ocean wanderer for at least two years before returning occasionally to its original birthplace in order to prospect for a future territory of its own, in readiness for breeding when it is about six or seven years old.

MANTA RAY

Despite its fearsome appearance and enormous size the manta ray is a harmless animal. It does possess teeth, but if feeds on nothing larger than shrimps and small fish. Many of its smaller relatives among the 340 species of skates and rays carry poison darts or electric shocks at the ends of their tapering or whip-shaped tails, but the manta does not. It is, however, a very strong animal and if accidentally hooked has been known to pull sizeable fishing boats through the water for several kilometres.

Also known as the 'devilfish', for no better reason than its appearance, the manta ray is a common animal throughout the tropics. Found wherever there are great numbers of its plankton food (along coasts as well as in the open oceans) it is the largest living ray, often weighing over 1.5 tonnes (tons) and measuring as much as seven metres (23 feet) across. Mantas often feed in pairs or small groups, using their 'wings' (which are enlarged fins) to glide effortlessly through the water in much the same way as a hang-glider uses his wings for support in the air. Unlike most rays, the manta's mouth extends across the front of its body. Two large mobile fins, one on either side, are known as 'horns' or 'head fins' and are believed to be used as scoops or as a funnel to lead food to the mouth. All the ray has to do is to 'fly' through shoals of plankton and small fish, flapping its wings like a bird or simply sending ripples down them for slower speeds, and funnel the food into its wide-open mouth.

Being a surface-dweller, the manta is more often seen than other skates and rays which spend the majority of their time lying on the ocean bed. It also makes its presence known with spectacular leaps out of the water, sometimes two metres (over six feet) clear above the surface. The manta ray sails through the air looking rather like a giant bat before hitting the sea again with a resounding boom.

DATA
SPECIES
Manta ray (*Manta birostris*)
CLASSIFICATION
Batoidea (skates and rays)
DISTRIBUTION
Mostly Atlantic tropical seas
HABITAT
Surface waters of open oceans
SIZE
Up to 7 metres (23ft) across; weighs nearly 1.5 tonnes (tons)
FOOD
Fish, small crustaceans and plankton

TOWNS AND SUBURBS

A few years ago the citizens of Miami, Florida, had some unexpected visitors. Alligators began to move into town in appreciable numbers and were turning up in people's swimming pools, on golf courses and even under cars.

Meanwhile, several thousand kilometres north, in a number of towns in sub-arctic Canada, polar bears began combing the local rubbish tips for food. There are now so many bears that the local people have to be driven to work in special buses to avoid encounters with the hungry animals.

All over the world, surprising numbers of animals and plants have learnt to tolerate the presence of people and are making homes for themselves in towns. Many species have even come to prefer their new existence in the concrete jungles scattered across the globe. Some, such as the alligators and polar bears, are sizeable enough to make their presence known, but few people realize how many other, less obvious, creatures share our towns and cities. Some are small, some are shy, others only come out at night. But depending on where in the world you live, you could unwittingly be sharing your neighbourhood with a multitude of beetles, bats, swifts, mice, moths, spiders, scorpions, geckos, silver-fish, mouse lemurs, racoons, koalas or puff adders. Some are more welcome than others and we attract these with nest boxes or scraps of food; with others we have waged all-out war.

In terms of evolutionary time, towns and cities have been around for only a very short period. Long before humans came on the scene, giant dinosaurs and other animals roamed the primaeval forests and swamps where enormous metropolises now stand. Large numbers of fossil sabre-toothed cats and other prehistoric creatures have been found under Los Angeles, while the remains of mammoths and woolly rhinos have been discovered beneath the streets of London.

Man has been the intruder ever since the first mud-hut towns sprang up in the Middle East some 10,000 years ago. Those first 'urban agglomerations' may not have been terribly alien to the plants and animals of the time, but over the years our cities have changed dramatically and become less and less hospitable to wildlife. In the Tokyo-Yokohama Metropolitan Area in Japan over 30 million people live in a vast conurbation covering some 2,800 square kilometres (1,080 square miles). Wildlife has had to adapt to this new habitat of concrete, tarmac, houses and high-rise blocks as best it can. Today, the world's population is estimated to be nearly 4,500 million. On average this means that there are roughly 30

Rosebay Willowherb *above*
With its attractive flowers, this plant grows in many city areas, quickly colonizing vacant lots.
Manhattan *right*
The familiar skyline of New York City looks an unlikely habitat for wildlife. But the city plays host to a wide variety of species.

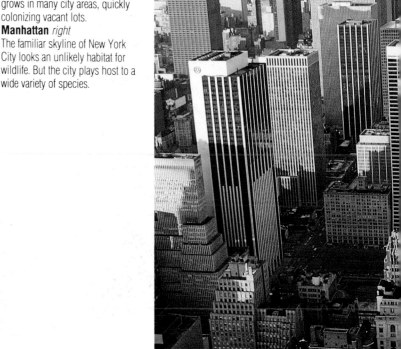

people per square kilometre (75 people per square mile) – but large areas are either too cold or too dry to support much life, so people are crowded together in the suitable places that are left. Increasingly, as the human population rises, wildlife has no choice but to fit in or die out.

Incredible though it may seem, wildlife is adapting. From New York City to a tiny house on an island in the South Pacific; from eskimo towns in the Arctic to Nairobi, almost sitting on the equator; from Lhasa, in Tibet, a small town 3,684 metres (12,000 feet) above sea level, to Ein Bokek, a settlement on the shores of the Dead Sea, 393.5 metres (1,300 feet) below sea level; wildlife has found a home.

Whole books have been written on the wildlife of London, New York and other major cities around the world. Even in Hong Kong, one of the world's most crowded places, where thousands of people have to live on boats because there is not enough space on the land, there is still a wealth of wildlife. Badgers have set up home on Wimbledon Common in the southern suburbs of London, and in people's gardens in Bristol; even peregrine falcons have been recorded hunting pigeons around the Tower of London.

Normally it is the patchwork of gardens and parks, leafy squares, churchyards and allotments in any town or city which attracts the wildlife. Such areas provide little oases amid the desert of concrete. But even the buildings themselves provide suitable habitats for a multitude of creatures. Their vertical faces sometimes offer the same nesting opportunities as natural cliff faces for species like the kestrel and that classic city dweller, the pigeon.

Inside the buildings themselves, many other species have successfully taken up residence. The house mouse joined us soon after we built our first settlements and has been our constant, albeit unwelcome, companion ever since. One species of termite has developed a taste for plastic, often stripping cables and causing major electrical faults, and of course there are the infamous beetle grubs we call 'woodworms' which burrow into furniture. In many parts of the Far East and Africa domestication has gone so far that some people do not consider their house a home unless a friendly gecko is in residence.

But the fact that some wild species have adapted in such extraordinary ways should not be an excuse for ignoring the spread of urbanization and its effects on the natural world. Sitting in a sea of asphalt and concrete, with thousands of people driving to and fro every day and filling the air with exhaust fumes, towns and suburbs could never be a true replacement for the world's natural habitats. But the wildlife has no choice.

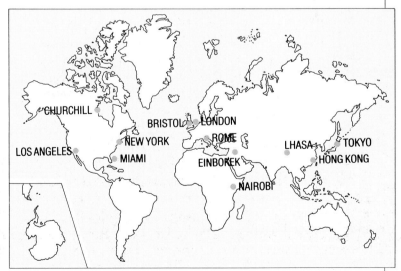

The world's cities *above*
The map pinpoints most of the towns and cities mentioned in this chapter.

145

Unwelcome guests

Many forms of wildlife that have made homes for themselves in towns and suburbs are most unwelcome visitors and residents. Snakes in particular are generally feared and disliked, especially when they come into houses. Some species, like the puff adder (*Bitis arietans*) which is a common suburban inhabitant in many parts of Africa, are a genuine cause for concern. A bite from one of these may kill a young child or a pet. But their reputation extends to all species, including the harmless ones. House snakes (*Boaedon* sp.) are occasionally tolerated because they prey on mice and other unwelcome guests, but even they have long teeth and therefore an extremely painful, albeit non-venomous, bite.

Fisk's house snake (*Lamprophis fiskii*) *above*
Common or brown rat (*Rattus norvegicus*) *top right*
Oriental cockroach (*Blatta orientalis*) *centre right*

In contrast, the brown rat (*Rattus norvegicus*) is unwelcome not so much because it bites, but because it causes tremendous damage to crops and stored food and because it transmits many diseases. Although it is a comparative newcomer to Europe (it arrived about 250 years ago) it is one of the most historic examples of wildlife in the city. It has more than made up for lost time, largely thanks to man's wasteful and unhygienic habits and in spite of all attempts to kill it. If man ever sets up home on the moon, rats will not be far behind.

Brown rats spend most of their time in cellars, basements, sewers and rubbish dumps. In houses they are generally secretive and discreet animals, so many people are unaware of their existence.

Another inveterate traveller, which has turned up virtually all over the world, is the cockroach. There are about 3,500 species of these insects, which have a reputation for being greasy and smelly. They are ideally suited to living in houses, even among people who are unfriendly towards them, because they have highly sensitive warning devices in their legs. These sensors register the slightest tremor as someone walks nearby, so the cockroach can scuttle away at the vaguest hint of danger. They have wings of a sort but rarely need to fly since they are probably the swiftest runners in the insect world, ready to escape at a moment's notice.

Cockroaches are most at home in kitchens, bathrooms and restaurants. The food they do not eat themselves they ruin with their horrible smell, and they are also able to pre-taste likely-looking food before actually eating it – just in case a householder has ideas about poisoning them.

House fly

The house fly (*Musca domestica*) has a shorter lifespan than any other insect. The females live on average just 29 days, while the males die after only 17 days. Females lay their eggs wherever the larvae will be able to find plenty of food – in piles of dung and rotting household rubbish, for example. The larvae hatch after only 12 to 24 hours and pupate within a few days. A week later the adult flies emerge and the females are ready to lay their eggs only a few days after that.

The house fly is also potentially the most dangerous insect in the world. It can transmit to humans 30 or more diseases and parasitic worms, including cholera, typhoid, dysentery, bubonic plague, leprosy and diphtheria.

Silverfish

The silverfish (*Lepisma saccharina*) is found in houses, apartments and other buildings – and virtually nowhere else. It is a scavenging insect, feeding on flour and other starch-rich food. Coated in shiny, metallic-looking scales, it is wingless and spends most of its time in damp, cool places such as cellars and baths. Apart from their size the young are almost identical to the adults, which are about one centimetre (two-fifths of an inch) long and regularly shed and re-grow their outer skeletons throughout life.

Tree frog

Widely distributed throughout tropical and temperate parts of the world, tree frogs are highly specialized amphibians adapted to an arboreal (tree-dwelling) way of life. They have such an extraordinary clinging ability, thanks to their toe pads, that they seem to defy gravity. They like to be close to water and some species seem to be as happy sitting in an empty bath or on a kitchen tap as on a dangling leaf in the jungle. The large, bulging eyes give them a wide-angle view and a certain amount of binocular vision, for gauging distances and ambushing passing insects – and for spotting potentially dangerous humans before they get too close.

Tree frog (*Rhacophorus leucomystax*)
One of the species common in kitchens and bathrooms.

COMMON RACOON

DATA

SPECIES
Common racoon
(*Procyon lotor*)

CLASSIFICATION
Carnivora (carnivores)

DISTRIBUTION
Southern Canada, USA, Central America

HABITAT
Diverse habitats, often near water; common in towns and cities

SIZE
Head and body length 60cm (2ft); weight up to 15kg (33lb)

FOOD
Virtually anything edible, including crayfish, frogs, fish, fruit, nuts and dustbin scraps

The unmistakable masked face and bushy, ringed tail of the racoon (*Procyon lotor*) is a familiar sight in many North American towns. Mischievous, insatiably curious, often destructive, it has become a notorious dustbin-raider from southern Canada across the USA right into Central America.

Racoons belong to a diverse group of animals which includes coatis, kinkajous and, depending on which scientists you believe, red and giant pandas as well. Distantly related to dogs, they are highly adaptable, equally at home living among people as in their natural habitats. They will sleep anywhere that seems safe, often in places like garden sheds and attics. Indeed, it is not uncommon for racoons to forget their nocturnal way of life and spend their days by the side of a road, begging food from passers-by.

Outside towns and cities, where there are no dustbins or roadsides, racoons like to forage near streams or marshy areas. They search for crayfish, frogs and fish, using their hands almost as skilfully as a monkey does, to catch and manipulate their food.

'Coons', as they are affectionately known, are not true hibernators. In the north of their range they do become inactive during winter, remaining in their dens for a month or more at a time if night temperatures do not rise above freezing, but they do not fall into the trance-like state of hibernation. As soon as the weather begins to improve, around early February, they mate. Up to seven young are born two months later, initially without the dark masks on their faces or the rings on their tails; both of these appear within a few days. The young coons start exploring their surroundings some time in June and soon go everywhere with their mother, keeping in touch in the dark with purrs, twitters and growls.

The long, freezing winter months which follow are hard on the young racoons. Those that do not starve are likely to lose more than half their body weight. Winter hardships, however, are only part of the problem. In the USA over four million coons are killed each year, either by hunting with specially-bred hounds, or by trapping. It is not surprising that, despite the risk of being killed by cars, racoons find it safer in urban areas, where so many of them have opted for a relatively easy and peaceful existence.

RED FOX

The red fox is the most widely distributed and probably the most adaptable of all carnivores. Its varied diet includes small deer, rabbits, mice, birds, beetles, grasshoppers, earthworms, blackberries, apples, dustbin scraps and bird food. Being a true opportunist, it will hide away its less favourite prey just in case it is needed in the future – and it nearly always remembers the locations of these stores.

Disliked by some suburban dwellers yet welcomed by others, foxes are common in towns and cities throughout their range. They even make their dens, or 'earths', under garden sheds or in the middle of dense flower beds. In these unlikely homes up to eight cubs are born in the spring. The female, or vixen, is in attendance all day and much of the night for the first three weeks, after which she lies up elsewhere.

As they grow older the cubs tentatively emerge from the earth to explore and play. By autumn they are confidently running around together, sometimes in broad daylight. They learn various skills for later in life, particularly how to catch rodents with the fox's characteristic 'mouse leap', springing a metre (three feet) or more off the ground before diving on to the prey.

Adult foxes are mostly solitary animals but they have an extremely complex social behaviour. The dominant individuals monopolize the best habitat and, while their paths may cross with others many times each night, they tend to avoid one another. They communicate by means of sounds (mostly yapping, barking, howling and whimpering) as well as scent-marking and visual signals with tail, face and ears. Their senses are excellent. Indeed, their hearing is so good that they catch worms by criss-crossing pastures after it has rained, listening for the rasping of the worms' bristles on the grass.

Millions of foxes have been slaughtered in recent years because they are important carriers of rabies. However, all attempts to rid them from town and country have so far been unsuccessful because they are such resilient animals. Three out of every four foxes in a population can be killed and still they will bounce back to their former numbers in a few years. The best hope for eliminating rabies in foxes now seems to lie in an oral vaccination, which would make them immune to the disease. If the majority of foxes in an area could be treated in this way the disease would then die out.

DATA
SPECIES
Red fox (*Vulpes vulpes*)
CLASSIFICATION
Carnivora (carnivores)
DISTRIBUTION
North America, Europe, many parts of Asia, North Africa
HABITAT
Diverse habitats including woodlands, hills and mountains, towns
SIZE
Head and body length 67cm (2ft 2in), tail length 41cm (1ft 4in); weight 7kg (15lb)
FOOD
Wide range, including small animals such as mice and rabbits; skilful scavenger in towns and cities

LEOPARD

ENDANGERED

The leopard (*Panthera pardus*) is the most widespread member of the cat family, found over much of Africa south of the Sahara and in many parts of southern Asia. It is a highly adaptable animal, able to live in a range of different habitats from tropical rain forests and arid savannahs to cold mountains or city suburbs. It even lives in urban areas such as Nairobi, in Kenya, where it sunbathes on house roofs and hunts in the gardens.

Nevertheless, leopard numbers have declined almost everywhere, largely due to hunting for their highly-prized fur and persecution after attacks on domestic animals. There are thought to be about 100,000 leopards left in the wild.

PIPISTRELLE BAT

DATA
SPECIES Pipistrelle bat (*Pipistrellus pipistrellus*)
CLASSIFICATION Chiroptera (bats)
DISTRIBUTION Widespread throughout Europe and east into India
HABITAT All types of habitat except very exposed regions
SIZE Head and body 35 to 45mm (1¼ to 1¾in), wingspan 190 to 250mm (7½ to 10in)
FOOD Small flying insects

There are nearly 1,000 different species of bats, representing nearly a quarter of the world's mammal species. They are found almost everywhere, except at the poles and on some of the highest mountains, and they range in size from the tiny bumblebee bat, weighing just 1.5 grams (one-twentieth of an ounce) and with a wingspan of 15 centimetres (six inches) to flying foxes which weigh a thousand times as much and have a wingspan of two metres (six feet) or more.

The pipistrelle (*Pipistrellus pipistrellus*) is the smallest native British bat. It is found in all kinds of habitats throughout much of Europe, east to Kashmir and north as far as the Arctic Circle. A nocturnal animal, it feeds mostly on small insects which are caught and eaten in flight. Flying at night poses many problems, not least of which are avoiding obstacles and locating food. Like most bats, the pipistrelle uses 'echolocation' to find its way. It produces complex high-frequency squeaks which bounce off objects and are picked up by its highly acute hearing.

During feeding flights the pipistrelle patrols a regular 'beat', uttering its shrill squeaks and taking small beetles, gnats, moths and other flying insects.

Although its food items consist of up to two-thirds water, pipistrelles also need to drink. They do this in two ways. One is to skim over a pond or stream like a small bird, taking sips from the surface. The second is actually to land near a pond or puddle and crawl like a slow, ungainly mouse to the water's edge.

During the summer, female pipistrelles form large nursing colonies. These are often in buildings and may contain up to 1,000 mothers. The single young are born from the third week in June until the second week in July and, within a very short time, begin to join the adults on their night-time feeding forays. The colony members emerge about half an hour after sunset, the young ones appearing first after a great deal of squeaking and shuffling. They fly intermittently during the night, often over or near water, and return around sunrise.

The males, meanwhile, are usually solitary or live in small groups. They normally choose to roost in old buildings or other confined spaces, but it is not uncommon to find them fast asleep on tree trunks or walls. All bat species, of both sexes, spend much of their time at these roosts washing and grooming, often hanging by one foot while the other vigorously combs all parts of the body.

Most of the winter is spent hidden in their chosen building or hollow tree, hibernating. They often wake up and take flights during mild weather, however, especially during the warmer days of late autumn and early spring. A temperature of around 5°C (40°F) is sufficient to rouse them from their deep sleep and send them out for an hour's foraging. Afterwards the bat may return to its former hibernating site, called a hibernaculum, or choose a new one. A damp place is often preferred so that the mammal does not become dehydrated during its sleep; often, hibernating individuals are covered with beads of moisture which are not wasteful perspiration but helpful condensation.

SKUNK

Looking like a cross between a badger and a weasel, the skunk is a common inhabitant of American urban areas. It is best known for its smelly defence mechanism and its role as a transmitter of the disease rabies and so, despite its appealing looks, it is not a very popular animal.

There are 13 different species of skunk, found all over North, Central and South America. They all have spectacular, though different, black and white coats which help to advertise their presence to intruders. If an adversary comes too close they wave their bushy tails in the air, stamp their front feet and walk around stiff-legged to warn that they might expel a fine spray of foul-smelling liquid. They can do this from less than a month old, spraying as far as seven metres (22 feet) away. The spray, an oily, yellow liquid produced by two glands under the skunk's tail, is aimed at the intruder's face. It causes intense skin irritation and even temporary blindness if it enters the eyes. For days afterwards the odour from this bitter-smelling mist persists, and

if it hits a person's clothes the stink hardly ever comes out.

Because of its scent the skunk has little to fear and usually trots around at a leisurely, unconcerned pace. Only the great horned owl ignores its smell and attacks this otherwise carefree animal.

Skunks forage at night, mostly for insects and small animals, grubs, birds' eggs and fruit, depending on the time of year. However, rabid individuals will attack virtually anything that moves and such animals are feared over much of the USA. As far as is known, skunk spray does not carry the rabies virus.

In their natural forest habitat, skunks use the discarded underground burrows of other species such as foxes, racoons and coyotes. Several may huddle together in the same den in winter, for a few days or weeks at a time, during severe weather. In towns, however, they have adapted well and often set up home in old buildings and outhouses. The skunk is a versatile creature, and a man-made environment seems to suit it.

DATA

SPECIES
Common or striped skunk
(*Mephitis mephitis*)

CLASSIFICATION
Carnivora (carnivores)

DISTRIBUTION
Southern Canada across the USA
to northern Mexico

HABITAT
Woods, grasslands, deserts,
towns and variety of other
habitats

SIZE
Head and body length 28 to
38cm (11 to 15in), tail length 18
to 43cm (7 to 17in); weight up to
2.5kg (5½lb)

FOOD
Mice, insects and other small
animals, fruit, grains and green
vegetation

WHITE STORK

DATA

SPECIES
White or European stork
(*Ciconia ciconia*)

CLASSIFICATION
Ciconiiformes (herons, storks
and flamingos)

DISTRIBUTION
Europe, Africa and many parts of
Asia

HABITAT
Marshes, wet meadows and
grassy plains; often nests near
towns or farms

SIZE
Height 100cm (3ft 3in)

FOOD
Insects such as grasshoppers,
earthworms, mice, lizards,
snakes, birds and other animals

Every spring, towns and villages all over Europe await the arrival of the white stork from Africa. An imposing bird, with wings that span nearly one and a half metres (five feet), it has always been looked upon with affection because of the old legend about it bringing babies. It also shows great fidelity to its nest site: year after year, the same storks return to their favourite chimney stacks and roofs to build their nests, and this has appealed greatly to local people over the centuries.

White storks are normally silent birds apart from hissing when annoyed, but they thoroughly enjoy clacking their bright-red bills together. The noise is so loud that it can be heard even before the birds are in sight. As they soar high above a town or village, floating on the thermals, the bill-clacking is the first indication that they have arrived.

People in many towns erect special high platforms, such as old cartwheels, to help the birds find a suitable base for the enormous nests they like to build. There is such a mass of nesting material that house sparrows often build their nests inside – sometimes there are a dozen or more sparrows' homes in a single stork nest.

Storks are also popular birds because they pay their rent by acting as scavengers. Their diet during nesting is varied, including scraps, rodents, fish, frogs, insects and earthworms. On their African winter grounds they are known as 'grasshopper birds' because they follow the locust swarms.

For all these reasons, the white stork has been protected in Europe for centuries. But between 1900 and 1958, populations decreased in numbers by as much as 80 per cent. Today they no longer nest in some countries, notably Sweden and Switzerland, and their numbers have dwindled to a few dozen in many others. The cause of their decline is not certain but there are probably many different reasons, including hunting in Africa, pesticide poisoning and changing agricultural practices. They are therefore no longer the welcome and reliable visitors they used to be – through no fault of their own.

TAWNY OWL

The nocturnal tawny owl (*Strix aluco*) is a familiar bird in many towns and cities over much of Europe and parts of Asia. Occasionally it can be seen during the daytime, hunched up in its tree roost in a park or large garden, being mobbed by smaller birds. But it is more often encountered at night, when its characteristic hooting carries even above the noise of traffic. Young birds, particularly in late summer and autumn, also make a sharp-sounding 'keewick' call. It is the combination of this and the adult's hoot which has led to the traditional, but completely inaccurate, description of the tawny owl's call as 'tu-whit, tu-whoo'.

GREY HERON

Standing motionless in the shallows of a pond or lake, the grey heron can easily be mistaken for a dead branch or post. Even when walking, its movements are slow and stealthy as it searches for fish, frogs and other small animals at the water's edge. Only when its neck, normally coiled into an S-shape, shoots out with startling speed when the prey is spotted, and its beak seizes the victim, is its presence given away.

Grey herons are highly specialized predators, highly adapted for capturing aquatic animals. When they fly, however, the special hunting features become cumbersome and awkward. They have to tuck in their long necks and let their legs trail out behind them as they go.

The grey heron is the best known and most widespread heron in the Old World. Despite some persecution and loss of habitat (due to drainage) it still maintains sizeable populations in most European countries. Hard winters often cause dramatic declines but even after these, numbers are regained in two or three years.

These ungainly birds nearly always nest in trees, as large groups in a colony known as a heronry. Most heronries these days consist of less than 200 occupied nests, but several years ago there was one in Holland which contained a record 1,000. There are smaller heronries in many city parks, such as Regent's Park in London. The birds from these often appear at garden ponds and take all the goldfish.

The first grey heron eggs are normally laid in March. An average clutch contains between three and five, which both parents incubate for nearly four weeks. When the youngsters are old enough they fly off in all directions, looking for new homes. Those in the extreme north of the herons' range tend to migrate southwards in the colder months with their parents.

In some parts of the world, in great lakes and marshes, herons join huge colonies of mixed water birds, including egrets, storks, ibises and spoonbills. The resulting flock, with its ceaseless activity and chatter, is an unforgettable sight.

DATA
SPECIES
Grey heron
(*Ardea cinerea*)
CLASSIFICATION
Ciconiiformes (herons, storks
and flamingos)
DISTRIBUTION
Throughout Europe, much of
Asia and Africa
HABITAT
Lakes, seashores, lagoons, rivers,
swamps and other areas of open
water – even garden ponds
SIZE
Height 90cm (3ft)
FOOD
Fish, larvae of diving beetles and
dragonflies, mice and voles

ELEPHANT HAWK MOTH

Although there are many more moths than butterflies, the former are probably less familiar to the town-dweller. Active during the night, in general they have no need of the bright, colours of butterflies and so are less conspicuous.

But some species of moths are colourful and dramatic, and are surprisingly common in suburban gardens and on rough waste ground. The spectacular elephant hawk moth is among these. Its caterpillars like to eat rosebay willow herb and other wild flowers which are common in urban areas, even in places like central London. The caterpillars look rather like legless lizards: dark brown in colour, they have six eye spots on the front of their abdomens and, when disturbed, withdraw their heads and puff out this region in a threatening posture.

The adult elephant hawk moth is only seen on the wing during the summer, mostly in June. It spends the rest of the year as an egg, caterpillar or chrysalis. As with all moths and butterflies, the adult is designed principally for flight and its main purpose in life is to find a mate, to ensure that another generation is produced. Flight requires a great deal of energy so the insects feed on sugary nectar, hidden deep inside flowers, which is an energy-rich source of food. The flowers also benefit because as the butterflies and moths feed they are dusted with pollen which will rub on to the next flower they visit. This fertilizes the flower and, in turn, ensures that another of its generations is produced.

If the moths simply used this high-energy fuel to fly around haphazardly, the chances of meeting a member of the opposite sex within their short lifespans would not be very great. So they have devised a special system whereby the female emits an airborne scent which is detected by the male's feathery antennae. A female releasing this scent, or pheromone, may draw in a number of males from a radius of several hundred metres, since the males are able to detect incredibly minute quantities of the scent in the air and accurately pinpoint its source. Soon after they have mated the females lay their eggs and the adults end their lives.

TOKAY GECKO

Geckos are the noisiest reptiles in the world. Most other lizards and snakes are silent, or only hiss, but geckos make extensive use of their voices. Their vocabulary ranges from quiet chirping and squeaking to a surprisingly loud, weird barking.

But the gecko's greatest claim to fame is as an insect-eating house guest. Many species have, to a great extent, lost their fear of people and inhabit houses in many parts of the world. They are welcomed nearly everywhere they go. In parts of Africa a house is not a home without a resident gecko; in Bangkok, it is reckoned to be an especially good sign if a gecko happens to be uttering its cry when a baby is born; and in many places the friendly little reptiles are trained to come regularly to a dining table and accept crumbs offered to them.

Most of a gecko's time is spent hanging on walls or upside down on ceilings. The way they do this has been studied in detail in the tokay gecko of South-East Asia. It does not have special suction pads on its feet, as was once thought, but instead uses microscopic hooks, or hairs, which grow all over the undersides of its flattened toes. Rather like the bristles of a brush, these hooks catch in the tiniest irregularities of a surface. They even enable the geckos to run up vertical plates of glass.

Geckos are found in many different habitats, from deserts and rainforests to high mountains and, of course, human dwellings. Wherever they live, they are mostly nocturnal and eat a range of small creatures including spiders, beetles, butterflies, millipedes, crickets and cockroaches. They have large eyes and excellent night vision but tend to see only objects that move. If an insect keeps perfectly still, a hunting gecko will ignore it; but as soon as it moves, the gecko will leap and seize it in a flash.

DATA	
SPECIES	
Tokay gecko (*Gekko gecko*)	
CLASSIFICATION	
Squamata (lizards and snakes)	
DISTRIBUTION	
South-East Asia	
HABITAT	
Diverse range of habitats, including houses, factories and farm buildings	
SIZE	
Up to 35cm (14in)	
FOOD	
Insects and mice	

GLOSSARY

Adaptation
A feature of a living thing that helps it to survive in a particular environment, such as the fatty blubber layer beneath the skin of a seal that keeps it warm in icy waters

Animal
Technically, a living thing that eats or consumes its food and digests it internally (compare Fungus and Plant)

Aquatic
Living in water

Arboreal
Living in trees

Boreal
The vast tracts of coniferous forest that stretch across northern North America, northern Europe and northern Asia

Brackish
Partly salty water, as found in an estuary, where sea water and fresh water mix

Camouflage
Any form of disguise used by a living thing, such as blending in with the background or mimicking another animal or plant; it can be used by both predators and prey

Carnivore
An animal or plant that eats only meat, or mostly meat; there is also a group of animals called Carnivora that includes cats, dogs, bears, weasels and racoons

Colonize
To settle in a new region and live there permanently

Commensal
Feeding together (literally 'at the same table')

Competition
In nature, when two individuals or species need the same resource (such as food or living space) and so have to strive against each other to obtain it

Conservation
Generally, to protect and preserve something – so that wildlife conservation means protecting and preserving natural areas with their animals and plants; more specifically, to use animals, plants and other natural resources in such a way as to improve the quality of life for mankind – but without wasting or using up those resources

Crepuscular
Active during twilight (dawn and dusk)

Desertification
The increase in size of desert regions as a result of man's activities

Diurnal
active during daylight

Dormancy
A period of inactivity during bad conditions, such as very cold or very dry weather; during dormancy an animal's body temperature and pulse rate do not fall (compare Hibernation)

Ecological niche
A lifestyle that exactly fits the adaptations of a certain species

Ecology
The study of animals, plants and their natural surroundings, and how they affect each other

Ecosystem
A recognizable large region, such as a forest or the sea, including all the animals and plants living there plus their natural surroundings

Endangered
In danger of becoming extinct; a specific term used by conservation organizations (rare, vulnerable and threatened are similar terms); the IUCN's Red Data Books give the official list of endangered plants and animals; in this book the term is used more generally to cover any species that is endangered or threatened with extinction

Endemic
Native and restricted to a certain region or country

Environment
The surroundings and circumstances of a living thing, including the terrain, other animals and plants, climate amount of moisture, and so on

Evolution
The gradual change with time of living things to become better adapted to their environment, by the process of natural selection; since the environment is also changing, evolution is a continuing process

Extinct
No longer living, died out completely (usually applied to species); an animal or plant species is officially extinct if there have been no certain records of it for 50 years

Feral
A domestic animal that lives in the wild, such as the 'alley cats' in big cities

Fungus
Technically, a living thing such as a mushroom, toadstool or yeast that feeds on rotting plant and animal remains, digesting the remains externally and then absorbing the nutrients through its outer layer

Genus
A group in animal and plant classification, made up of one or more species

Gestation
The period of pregnancy, from conception to birth

Gregarious
Living with others of the same kind, such as zebras that live in herds; or starlings that live in flocks (compare Solitary)

Habitat
The type of place where an animal or plant lives, such as a wood, desert or seashore; usually, several habitats in the same region form an ecosystem

Herbivore
An animal that eats only, or mostly, plants

Hibernation
A period of inactivity during bad conditions, such as very cold weather; during hibernation an animal's body temperature, pulse rate and breathing rate all fall (compare Dormancy)

Home range
The area over which an individual animal habitually roams to find food

Invertebrate
An animal without a backbone

IUCN
International Union for the Conservation of Nature and Natural Resources

Larva
The immature form of an animal, which does not usually look like a grown-up; for example, the caterpillar is the larva of the butterfly

Marine
Living in the sea

Migration
A regular journey, usually linked to the seasons, when animals move from one place to another to feed or breed

Native
Originating from a certain region or country

Natural selection
The evolutionary process by which nature 'chooses' among the variety of living things, selecting those which are most suited to a particular environment; only the best adapted survive and produce offspring, hence the term 'survival of the fittest'

Nocturnal
Active during darkness

Omnivore
An animal that eats both plants and animals

Organism
A living thing, including all animals, plants, fungi, and microbes such as bacteria

Parasite
A living thing that obtains its nourishment or living place from another, harming it in the process

Pesticide
A chemical which kills pests – a 'pest' being any animal or plant that humans do not want, such as insects that damage crops and waterweeds that choke canals

Phytoplankton
Minute plants that float near the surface of a body of water

Plankton
See Phytoplankton, Zooplankton

Plant
Technically, a living thing that captures the energy in sunlight to live and grow

Population
A group of animals or plants of the same species which are fairly separate from other such groups

Predator
The hunter, an animal that kills other animals for food

Prey
The hunted, an animal that is killed and eaten by another animal

Range
The area over which a species is distributed, apart from odd stragglers that stray outside this area

Solitary
Living alone, such as the tiger; which only meets another tiger to mate (compare Gregarious)

Specialization
Becoming adapted to one feature of the environment, such as the giraffe, which specializes in eating leaves in the high branches

Species
A group of living things that look similar and can breed successfully

Sub-species
A sub-group within a species, which is recognizable and usually restricted to a certain area such as a lake or mountain range

Symbiosis
Living things from different species that help each other in some way

Terrestrial
Living on land

Territory
An area 'owned' by an animal or group of animals, which is defended against other members of the species

Vertebrate
An animal with a backbone

WWF
World Wildlife Fund

Zooplankton
Minute animals that float near the surface of a body of water

INDEX

The publishers wish to thank the following for their kind permission to reproduce the photographs in this book:
Bruce Coleman Ltd 19, 53-55, 57-61, 71, 75, 82-98, 100-12, 114 A and C, 115-43; **J and D Bartlett** 27, 49 L, 79 R, 80; **M Boulton/WWF** 67 R; **E Breene Jones** 37 R; **J Burton** 12 L, 24, 25 B, 64 B, 65 A, 149 L, 150, 153 L, 154; **B and C Calhoun** 40 R, 48, 151; **R I M Campbell/Gorilla Research** 18 C and R, 70; **A Compost** 21 A; **E Crichton** 32; **M Dakin** 35 A; **P Davey** 63 L; **G Doré** 23; **F Erize** 21 B, 51 L, 64 C, 74; **O Eshbol** 49 R; **M Fogden** 10 L, 13 L, 16, 17 A, 20, 42, 43 R, 50, 67 L, 155; **J Foott** 40 L, 45 R, 47, 65 B; **M Freeman** 13 C; **C B and D W Frith** 14, 15, 64 A, 147 BR; **D Green** 25 C, 153 R; **M P Harris** 76; **C Henneghien** 144-45; **P A Hinchcliffe** 28; **U Hirsch** 152; **C Hughes** 46; **H Jungius/WWF** 44 L; **M P Kahl** 78; **S J Krasemann** 34 R, 45 L; **G Langsbury** 79 L; **J Langsbury** 144 A; **W Lankinen** 29 L, 34 L, 36, 38; **L Lee Rue III** 33, 77; **L C Marigo** 13 A, 62; **J Markham** 147 L; **R K Murton** 37 L; **L M Myers** 72-3; **C Ott** 35 B, 73 A; **Prato** 10 R; **M Price** 149 R; **H Reinhard** 25 A, 26, 30, 35 C, 36 L, 69 L, 146 AR, 148; **F Sauer** 17 B; **E Schuhmacher/WWF** 81; **K Taylor** 41, 146 BR, 147 AR; **J Van Wormer** 44 R; **J Visser** 146 L; **K Weber/WWF** 29 R; **R Williams** 18 L, 68; **R Wilmshurst** 22, 31; **G Zeisler** 12 C, 39, 66, 69 R.
    Additional material: **B and C Alexander** 56; **H Angel/Biofotos** 99; **Impact Photos/A Le Garsneur** 114 B; **Oxford Scientific Films Ltd/ A Bannister** 51 R

Author's acknowledgements
Many people have helped with the production of this book. In particular, I would like to thank Philip Wilkinson, for his valued advice, infectious enthusiasm and unerring patience; Steve Parker, for his valuable contributions to and comments on the text; Jeremy Bratt and Hugh Schermuly for their work on the design of the book; Celina Dunlop for picture research; Maryann Rogers for co-ordinating printing and production services; Debra Taylor, for helping with yet more typing and re-typing; Ivan Hattingh for his much appreciated support and encouragement, as always; and Craig Johnson, for helping out in busy times.